Johannes-Robert Bruch

Parameter estimation for challenging phase equilibria

SCIENTIA
Das interdisziplinäre Wissenschaftsjournal
Sonderdruck 6 (2021)

Arbeitsgemeinschaft für internationale Beziehungen
Cooperation for international relations

Marosi Verlag, Dr. phil. Silvia Marosi

Bibliographische Information der Deutschen Nationalbibliothek
Die Deutsche Nationalbibliothek verzeichnet diese Publikation in der Deutschen Nationalbibliographie.
Detaillierte bibliographische Daten sind im Internet über http://dnb.d-nb.de abrufbar.

Dr. phil. Silvia Marosi, Brandenburger Straße 16, D-50668 Köln &
Arbeitsgemeinschaft für internationale Beziehungen
Cooperation for international relations
Unterheide 25, D-46519 Alpen (Niederrhein)

Druck: Books on Demand GmbH, In de Tarpen 42, D-22848 Norderstedt
Printed in Germany 2021

ISBN 978-3-945636-20-6
ISSN 2511-5227

Abstract

This thesis consists of three parts; in the first part, parameter sets of the Non-Random Two-Liquid (NRTL) model for five challenging binary liquid-liquid equilibria (LLE) relevant to the cluster of excellence Tailor-Made Fuels from Biomass (TMFB) are generated by means of the AVT.SVT in-house tool BOARPET (Bilevel Optimization Algorithm for Rigorous and Robust Parameter Estimation in Thermodynamics). In particular, to the author's best knowledge, this work is the first to report a model for the LLE of 2,5-DMTHF/water. In addition, parameter sets that allow modelling of the LLE of 2,5-DMF/water, 2-MTHF/water and butanone/water are regressed. The same measurement data used in BOARPET, are used to test the Data Regression System (DRS) in Aspen Plus V8.8, and the guarantees that both tools can give regarding thermodynamic stability are compared. It is found that for butanone/water, the regression results of Aspen Plus DRS violate thermodynamic criteria. First, the resulting model contains a spurious second phase split. In addition, at a certain temperature, the phase compositions that are predicted by Aspen Plus DRS based on this model do not fulfil the isoactivity condition.

In the 2nd part, the risks that may result from the aforementioned violation of thermodynamic criteria for process simulation are assessed, by means of several unit operation models in Aspen Plus, as well as flash simulations, formulated and solved in GAMS. The respective parameter sets are selected from the first part, either because these are found to result in violation of thermodynamics, or due to the particular shape of the implied Gibbs free energy curve with respect to composition. It is found that *i)* regression results violating thermodynamics can lead to flash results far off experimental evidence and thus, can cause considerable errors in, e.g., process simulation. This risk is present irrespective of the capabilities of the flash solver (local, global). In addition, *ii)* in some cases, the unit operation models in Aspen Plus solve the flash only to local optimality, which is reason to conjecture that even in case of correct regression results, there is a risk of incorrect results in process simulation with Aspen Plus.

In the 3rd part, stoichiometric and non-stoichiometric formulations from literature for the simulation of combined diffusive and chemical equilibrium are compared using local and global solvers through GAMS, for the case study SBA/DSBE/water. As agreement between the results of RGIBBS in Aspen Plus and BARON in GAMS is good, to date, no conclusions can be drawn on the risks, potentially resulting from simulation of reactions in Aspen Plus.

Acknowledgement

The publication is made within the framework of a scientific cooperation at the Vilnius University of Technology in Lithuania.

Contents

Nomenclature

Table 0.1.: List of abbreviations.

Abbreviation	Description
2,5-DMF	2,5-dimethylfuran
2,5-DMTHF	mixture of cis-2,5-dimethyltetrahydrofuran and trans-2,5-dimethyltetrahydrofuran
2-MF	2-methylfuran
2-MTHF	2-methyltetrahydrofuran
BOARPET	Bilevel Optimization Algorithm for Rigorous & Robust Parameter Estimation in Thermodynamics
BARON	Branch-And-Reduce Optimization Navigator
CPU	central processing unit
DECHEMA	DECHEMA Gesellschaft für Chemische Technik und Biotechnologie e.V.
DETHERM-SDC	DETHERM-Stoffdatencompiler
DPP	DECHEMA Data Preparation Package
DRS	Data Regression System
DSBE	di-sec-butyl ether
EoS	equation of state model
GAMS	General Algebraic Modeling System
KKT	Karush-Kuhn-Tucker
LICQ	Linear independence constraint qualification
LLE	liquid-liquid equilibria
llp	lower-level problem
max	maximum
min	minimum
NLP	Non-linear problem
NRTL	Non-random two-liquid model
Ref.	Reference
SBA	sec-butanol
TMFB	Tailor-Made Fuels from Biomass
ulp	upper-level problem
UNIQUAC	Universal quasi-chemical activity coefficient model
VLE	vapour-liquid equilibria
VLLE	vapour-liquid-liquid equilibria

Table 0.2.: List of indices.

Index	Description
aq	aqueous
E	excess
e	element
est	estimated value in regression
exp	experimentally-measured value in regression
f	formation
org	organic
i	constraints
i	species
$ideal$	ideal mixing
$iter$	index of the variable of Formulation 5.3
j	phase
k	experiment
l	lower level variable
M	mixing
ref	reference state
rx	example reaction
sc	scaling
u	upper level variable
∞	infinite dilution

Table 0.3.: List of symbols.

Symbol	Description
α	non-randomness parameter of the NRTL model
γ	activity coefficient
Δ	difference
$\boldsymbol{\lambda}$	vector of Lagrange multipliers λ
λ	Lagrange multiplier
ξ	conversion
μ	chemical potential
ν	stoichiometric coefficient
τ	binary interaction parameter
$\boldsymbol{\phi}$	vector of molar phase fractions ϕ
ϕ	molar phase fraction
φ	fugacity coefficient
φ_i^0	fugacity coefficient of species i at saturation pressure

Table 0.3.: List of symbols. (continued)

Symbol	Description
$\boldsymbol{\psi}$	vector of Lagrange multipliers for inequality constraints ψ
ψ	Lagrange multiplier for inequality constraints
$\mathbf{A}$	substance formula matrix
A	coefficient of extended Antoine equation
$A_{ij}^{BOARPET}$	NRTL parameter for species i and j in BOARPET
a_{ij}^{Aspen}	NRTL parameter for species i and j in Aspen Plus
$\mathbf{b}$	elemental-abundance vector
$B_{ij}^{BOARPET}$	NRTL parameter for species i and j in BOARPET
b_{ij}^{Aspen}	NRTL parameter for species i and j in Aspen Plus
$\mathbf{c}$	covariance matrix of the residual vector
C	heat capacity coefficient
$C_{ij}^{BOARPET}$	NRTL parameter for species i and j in BOARPET
$C_{p,i}$	heat capacity
c_i	constraint with index $i \in E \cup I$
c_{ij}^{Aspen}	symmetric NRTL parameter for species i and j in Aspen Plus
$D_{ij}^{BOARPET}$	NRTL parameter for species i and j in BOARPET
d_i	distribution ratio
d_{ij}^{Aspen}	symmetric NRTL parameter for species i and j in Aspen Plus
$\mathcal{E}$	index set of equality constraints
$E_{ij}^{BOARPET}$	symmetric NRTL parameter for species i and j in BOARPET
e_{ij}^{Aspen}	NRTL parameter for species i and j in Aspen Plus
F	set of feed compositions
$F_{ij}^{BOARPET}$	symmetric NRTL parameter for species i and j in BOARPET
$F_{P,i}$	Poynting factor
f	objective function
f_i	fugacity of species i
f_{ij}^{Aspen}	NRTL parameter for species i and j in Aspen Plus
G	molar Gibbs free energy
g	reduced molar Gibbs free energy
$\mathbf{g}$	vector of inequality constraints
ΔH_{rx}^0	enthalpy of reaction rx
$\mathbf{h}$	vector of equality constraints

Table 0.3.: List of symbols. (continued)

Symbol	Description
$\mathcal{I}$	index set of inequality constraints
I	index set of species $i \in 1, ..., N$
J	index set of phases $j \in 1, ..., M$
K	index set of experiments $j \in 1, ..., L$
M	molar mass
n	mole number
n_i^0	molar amount of species i in the feed
P	index set of parameter vectors $\mathbf{p}$
$\mathbf{p}$	vector of regressed parameters
p	pressure
p^0	reference pressure
p_i^s	saturation pressure of species i
R	ideal gas constant
R	number of independent reactions
r	residual vector
ΔS_{rx}^0	entropy of reaction rx
T	temperature
T^0	reference temperature
$\mathbf{t}$	vector of fixed variables in regression
$tran$	translation parameter for NRTL in BOARPET
u	standard uncertainty or standard deviation
$\mathbf{w}$	weighting matrix in Least Squares function
w	weight in objective function
$\mathbf{x}$	matrix of mole fractions or matrix of variables
X	index set of matrices of variables
x	mole fraction of a species in a homogeneous phase
x	regressed variable in regression
x_i^0	mole fraction of species i in the feed
∇	Nabla operator

List of Figures

List of Tables

1. Introduction

Valid models of phase equilibria are crucial for computer-aided process design. The two main classes of models applied to describe phase equilibria are excess Gibbs free energy (G^E) models and equation of state models (EoS). As the parameters of these models cannot be measured directly, they are usually estimated based on experimental data. Therefore, formulations of parameter estimation problems are required which guarantee thermodynamically correct results, including stability (8) and the same numbers of phases and phase splits as evident from experiments. State-of-the-art commercial parameter estimation tools such as Aspen Plus DRS (7) often implement regression problem formulations which do not consider thermodynamic stability rigorously. In order to contrast regression methods and to assess the risks which arise from using results from different regression tools within commercial process simulators, several methods of regression, as well as flash simulations, are compared in the present work.

The case studies considered in the present work are the binary liquid-liquid equilibria (LLE) of water with each of the following substances: Butanone, 2,5-dimethylfuran (2,5-DMF), 2,5-dimethyltetrahydrofuran (2,5-DMTHF), 2-methylfuran (2-MF), 2-methyltetrahydrofuran (2-MTHF). All these substances are considered conceivable fuel components or respective platform chemicals, from renewable sources in the literature (40; 66; 94; 39; 68; 41; 21). For instance, the study of binary LLE with water is relevant both for solvent extraction of fermentation products as well as for biofuels that will be exposed to air humidity. Further applications include utilisation as solvents for extraction processes (10; 50; 85).

In Chapter 3 of the present work, different formulations of parameter estimation problems for binary LLE employing G^E models are compared. Herein, special attention is paid to criteria for thermodynamic stability that are representing demanding constraints within the problem formulations.

In Chapter 4, regression results obtained with Aspen Plus DRS and the AVT.SVT in-house tool BOARPET (Bilevel Optimization Algorithm for Rigorous and Robust Parameter Estimation in Thermodynamics) for the aforementioned TMFB case studies are presented. In addition, a literature review regarding available measurement data and parameter sets for (G^E) models is given.

In Chapter 5, the risks resulting from i) incorrect regression results, as well as ii) local (as opposed to global) solvers, for process simulation are assessed. These studies are carried out in Aspen Plus and GAMS with up to three different formulations for the simulation of LLE at hand. Herein, the most challenging parameter configurations of the case studies from Chapter 4 in addition to examples from literature are employed.

In Chapter 6, stoichiometric and non-stoichiometric formulations from literature for the simulation of combined diffusive and chemical equilibrium are compared using local and global solvers through GAMS, as well as the unit operation models RGIBBS and REQUIL in Aspen Plus, for the case study SBA/DSBE/water.

Finally, in Chapter 7, a conclusion is presented summarising the main findings of the present work and giving an outlook on need for further research.

2. Fundamentals

In this chapter, an introduction into the mathematical and physical fundamentals of the present work is given.

2.1. Fundamentals of optimisation

A general formulation for constrained nonlinear optimisation problems is given by

$$\min_{\mathbf{x}\in\Re^n} f(\mathbf{x}) \tag{2.1}$$

subject to

$$c_i = 0, i \in \mathcal{E}, \tag{2.2}$$

$$c_i \geqslant 0, i \in \mathcal{I}, \tag{2.3}$$

where f and c_i are twice continuously differentiable, potentially nonlinear, functions. In this section the necessary and the sufficient conditions for local minima of problems of this form are discussed. To this end, the Lagrangian function reads

$$\mathcal{L}(\mathbf{x}, \boldsymbol{\lambda}) = f(\mathbf{x}) - \sum_{i\in\mathcal{E}\cup\mathcal{I}} \lambda_i c_i(\mathbf{x}) . \tag{2.4}$$

The problem formulations in the present work are based on the assumption that the linear independence constraint qualification (LICQ) holds for all minima. Under this assumption, the 1st-order Karush-Kuhn-Tucker (KKT) conditions are necessary conditions for a local minimum of a constrained nonlinear problem (NLP), see e.g. Nocedal and Wright (60). They can be formulated as follows: If $\mathbf{x}^*$ is a local minimum, there exists a unique Lagrange multiplier vector $\boldsymbol{\lambda}^*$ such that

$$\begin{aligned}\nabla_{\mathbf{x}}\mathcal{L}(\mathbf{x}^*,\boldsymbol{\lambda}^*) &= 0,\\ c_i(\mathbf{x}^*) &= 0, \text{ for all } i \in \mathcal{E},\\ c_i(\mathbf{x}^*) &\geqslant 0, \text{ for all } i \in \mathcal{I},\\ \lambda_i^* &\geqslant 0, \text{ for all } i \in \mathcal{I},\\ \lambda_i^* c_i(\mathbf{x}^*) &= 0, \text{ for all } i \in \mathcal{I}.\end{aligned} \tag{2.5}$$

For the compact formulation of 2nd-order conditions, further definitions are required. The active set $\mathcal{A}(\mathbf{x})$ consists of the indices of the equality constraints together with those of the inequality constraints for which $c_i(\mathbf{x}) = 0$. The critical cone is defined as follows (60)

$$\mathcal{C}(\mathbf{x}^*,\boldsymbol{\lambda}^*) = \left\{ \mathbf{w} \in \Re^n \middle| \begin{array}{ll} \mathbf{w}^T\nabla c_i(\mathbf{x}^*) = 0, & \text{for all } i \in \mathcal{E},\\ \mathbf{w}^T\nabla c_i(\mathbf{x}^*) \geqslant 0, & \text{for all } i \in \mathcal{A}(\mathbf{x}^*) \cap \mathcal{I} \text{ with } \lambda_i^* = 0,\\ \mathbf{w}^T\nabla c_i(\mathbf{x}^*) = 0, & \text{for all } i \in \mathcal{A}(\mathbf{x}^*) \cap \mathcal{I} \text{ with } \lambda_i^* > 0. \end{array} \right\} \tag{2.6}$$

Using these definitions, the 2nd-order necessary KKT conditions can be stated as follows (60). Suppose $\mathbf{x}^*$ is a local minimum and $\boldsymbol{\lambda}^*$ is the Lagrange multiplier vector for which the 1st-order KKT conditions are satisfied, then

$$\mathbf{w}^T\nabla_{xx}^2\mathcal{L}(\mathbf{x}^*,\boldsymbol{\lambda}^*)\,\mathbf{w} \geqslant 0, \text{ for all } \mathbf{w} \in \mathcal{C}(\mathbf{x}^*,\boldsymbol{\lambda}^*). \tag{2.7}$$

To obtain the 2nd-order sufficient KKT conditions Equation (2.7) has to be satisfied strictly for all $\mathbf{w}$ in the critical cone except the zero vector. Thus, x^* is a strict local minimum, if there is a $\boldsymbol{\lambda}^*$ for which the 1st-order KKT conditions are satisfied and the following condition is fulfilled

$$\mathbf{w}^T\nabla_{xx}^2\mathcal{L}(\mathbf{x}^*,\boldsymbol{\lambda}^*)\,\mathbf{w} > 0, \text{ for all } \mathbf{w} \in \mathcal{C}(\mathbf{x}^*,\boldsymbol{\lambda}^*), \mathbf{w} \neq 0. \tag{2.8}$$

2.2. Thermodynamic fundamentals

A necessary and sufficient condition for thermodynamic stability in a closed system at constant temperature and pressure is that the Gibbs free energy is globally minimal, (which is not to be confused with the necessary and sufficient conditions for a local minimum of a constrained nonlinear problem, see Section 2.1). Regarding multi-component phase equilibria (with the exception of pure phases), this is equivalent to a supporting hyperplane that fulfils Baker's criterion (8; 55). 2nd-order sufficient KKT conditions for local minima of Gibbs free energy (see Equation 2.8) are regarded as necessary (as opposed to sufficient) conditions for thermodynamic stability. The term "stability" used in the present work is excluding metastability. The necessary and sufficient criterion for thermodynamic stability in a closed system without reactions at constant temperature and pressure can be formulated as the following constrained nonlinear optimisation problem, see e.g. Walas (90)

$$\min_{\mathbf{x},\phi} \sum_{j \in J} \phi_j G(\mathbf{x}_j) \tag{2.9a}$$

$$\text{subject to}$$

$$\sum_{i \in I} x_{i,j} - 1 = 0, \text{ for all } j \in J, \tag{2.9b}$$

$$\sum_{j \in J} x_{i,j}\phi_j - x_i^0 = 0, \text{ for all } i \in I, \tag{2.9c}$$

$$\phi_j \geqslant 0, \text{ for all } j \in J, \tag{2.9d}$$

$$x_{i,j} \geqslant 0, \text{ for all } i \in I, j \in J. \tag{2.9e}$$

Herein, G denotes the molar Gibbs free energy. Moreover, ϕ_j is the molar phase fraction of phase j and element of the vector ϕ, while $x_{i,j}$ is the mole fraction of species i in phase j and element of $\mathbf{x}_j$, which is a column vector of the matrix $\mathbf{x}$. The mole fraction of species i in the feed is denoted by $x_i^0 \in [0, 1]$. $I = \{1, ..., N\}$ is the index set of species and $J = \{1, ..., M\}$ is the index set of phases. Equations (2.9a) - (2.9e) exemplify a formulation of a so-called flash problem that has to be solved to global optimality in order to determine the thermodynamically stable state of the specified system. The Lagrangian function for this problem can be stated as

$$\mathcal{L}(\mathbf{x}, \phi, \boldsymbol{\lambda}, \boldsymbol{\psi}) = \sum_{j \in J} \phi_j G(\mathbf{x}_j) - \sum_{j \in J} \lambda_j \left(\sum_{i \in I} x_{i,j} - 1 \right) - \sum_{i \in I} \lambda_{M+i} \sum_{j \in J} \left(x_{i,j}\phi_j - x_i^0 \right) - \sum_{j \in J} \sum_{i \in I} \psi_{i,j} x_{i,j} - \sum_{j \in J} \psi_{N+1,j} \phi_j, \tag{2.10}$$

where $\boldsymbol{\lambda}$ denotes the $(M + N) \times 1$ vector of Lagrange multipliers λ for equality constraints and $\boldsymbol{\psi}$ denotes the $(N + 1) \times M$ matrix of Lagrange multipliers ψ for inequality constraints.

The 1st-order KKT conditions (see Equation 2.5) for this problem read

$$\begin{aligned}
\nabla_{\mathbf{x},\boldsymbol{\phi}}\mathcal{L}\left(\mathbf{x}^{*},\boldsymbol{\phi}^{*},\boldsymbol{\lambda}^{*},\boldsymbol{\psi}^{*}\right) &= 0,\\
\sum_{i\in I} x_{i,j}^{*} - 1 &= 0, \text{ for all } j \in J,\\
\sum_{j\in J} x_{i,j}^{*}\phi_{j}^{*} - x_{i}^{0} &= 0, \text{ for all } i \in I,\\
x_{i,j}^{*} &\geqslant 0, \text{ for all } i \in I, j \in J,\\
\phi_{j}^{*} &\geqslant 0, \text{ for all } j \in J,\\
\psi^{*} &\geqslant 0, \text{ for all } \psi^{*} \in \boldsymbol{\psi}^{*},\\
\psi_{i,j}^{*}x_{i,j}^{*} &= 0, \text{ for all } i \in I, j \in J,\\
\psi_{N+1,j}^{*}\phi_{j}^{*} &= 0, \text{ for all } j \in J.
\end{aligned} \tag{2.11}$$

To demonstrate the meaning of stationarity of the Lagrangian function, the gradient of the Lagrangian function with respect to $\mathbf{x}$ and $\boldsymbol{\phi}$ is expressed for a binary system ($I = \{1,2\}$) with two phases ($J = \{1,2\}$):

$$\nabla_{\mathbf{x},\boldsymbol{\phi}}\mathcal{L}\left(\mathbf{x},\boldsymbol{\phi},\boldsymbol{\lambda},\boldsymbol{\psi}\right) = \begin{pmatrix} \phi_{1}\frac{\partial G}{\partial x}|_{x_{1,1}} - \lambda_{1} - \lambda_{3}\phi_{1} - \psi_{1,1}\\ \phi_{2}\frac{\partial G}{\partial x}|_{x_{1,2}} - \lambda_{2} - \lambda_{3}\phi_{2} - \psi_{1,2}\\ -\phi_{1}\frac{\partial G}{\partial x}|_{x_{1,1}} - \lambda_{1} - \lambda_{4}\phi_{1} - \psi_{2,1}\\ -\phi_{2}\frac{\partial G}{\partial x}|_{x_{1,2}} - \lambda_{2} - \lambda_{4}\phi_{2} - \psi_{2,2}\\ G\left(x_{1,1}\right) - \lambda_{3}x_{1,1} - \lambda_{4}x_{2,1} - \psi_{3,1}\\ G\left(x_{1,2}\right) - \lambda_{3}x_{1,2} - \lambda_{4}x_{2,2} - \psi_{3,2} \end{pmatrix}. \tag{2.12}$$

Equivalently to the flash problem given by Equations (2.9a) - (2.9e) the Baker's criterion is a necessary and sufficient criterion for thermodynamic stability in a closed system without reactions at constant temperature and pressure: A matrix of mole fraction variables $\mathbf{x}$ and a vector of phase fractions $\boldsymbol{\phi}$ fulfils thermodynamic stability, if and only if the common tangent hyperplane of the predicted phases does not lie above the hypersurface of Gibbs free energy at any admissible composition. The molar Gibbs free energy of a mixture is often expressed as a sum of the mole fractions x_i^0 times the molar Gibbs free energies G_i^{ref} defined at a particular reference state for species $i \in I$, plus the Gibbs free energy of mixing G^M:

$$G = \sum_{i\in I} x_i^0 G_i^{ref} + G^M. \tag{2.13}$$

As the second derivative of the first term in Equation 2.13 is zero, it does not have any influence on the compositions predicted for flash problem of systems without chemical or nuclear reactions and with the same reference state for all phases. A dimensionless measure for the Gibbs free energy is obtained by defining the reduced Gibbs free energy g using the molar thermal energy RT.

$$g = \frac{G}{RT}. \tag{2.14}$$

In the present work, the reduced Gibbs free energy of mixing $g^M(x)$ as a function of the composition is used for presentation in graphs and instead of the molar Gibbs free energy in the formulations in Chapter 5. The Gibbs free energy of mixing is usually split into an ideal part and an excess part:

$$G^M = G^{ideal} + G^E. \tag{2.15}$$

The ideal part can be derived from statistical mechanics using Stirling's approximation of the factorial, see e.g. Lucas (47):

$$G^{ideal} = RT \sum_{i \in I} x_i \ln(x_i). \tag{2.16}$$

The molar excess Gibbs free energy can be expressed as

$$G^E = RT \sum_{i \in I} x_i \ln(\gamma_i), \tag{2.17}$$

with γ_i being the activity coefficient of species i. The well-known condition of isopotential is given by

$$\mu_{i,\hat{j}} = \mu_{i,j}, \text{ for all } i \in I, j \in J, \hat{j} \in J. \tag{2.18}$$

Herein, $\mu_{i,j}$ denotes the chemical potential of species i in phase j. The chemical potential $\mu_{\hat{i},j}$ at constant temperature and pressure is defined as the partial derivative of the extensive Gibbs free energy $n_j G(\mathbf{n}_j)$ with respect to the mole number $n_{\hat{i},j}$ of species $\hat{i}$ in phase j at constant mole numbers $n_{i,j}$ for all $i \in \{I | i \neq \hat{i}\}$ (55)

$$\mu_{\hat{i},j} := \left. \frac{\partial n_j G(\mathbf{n}_j)}{\partial n_{\hat{i},j}} \right|_{p,T,n_{i \neq \hat{i},j}}. \tag{2.19}$$

Herein, n_j denotes the total mole number of the system and $\mathbf{n}_j$ denotes the vector of mole numbers of all species in set I in phase j. The isopotential condition (Equation (2.18)) can be derived from 1st-order KKT conditions (Equation 2.11) of minimisation of Gibbs free energy, see e.g. Smith et al. (74), but is not sufficient for 2nd-order KKT conditions. Therefore, it is necessary but not sufficient for stability. Using Equations (2.13) - (2.17) the chemical potential $\mu_{i,j}$ can be expressed as follows:

$$\mu_{i,j} = \mu_i^{ref} + RT \ln (x_{i,j}\gamma_{i,j}), \tag{2.20}$$

where μ_i^{ref} denotes the chemical potential of species i in a particular reference state. The product $x_{i,j}\gamma_{i,j}$ is denoted as activity $a_{i,j}$. Assuming that the same reference state is used for both phases, Equation (2.18) can be transformed into the so-called isoactivity condition using Equation (2.20)

$$a_{i,\hat{j}} = a_{i,j}, \text{ for all } i \in I, j \in J, \hat{j} \in J. \tag{2.21}$$

The activity coefficients $\gamma_{i,j}$ can be obtained from derivatives of the excess Gibbs free energy.

2.3. Non-Random Two-Liquid model

The equation that is employed in the present work to model the excess Gibbs free energy is the Non-Random Two Liquid model (NRTL) of Renon and Prausnitz (65). For the binary system with the species 1 and 2 this G^E model reads

$$\frac{G^E}{RT} = x_1 x_2 \left(\frac{\tau_{12} \exp(-\alpha\tau_{12})}{x_2 + x_1 \exp(-\alpha\tau_{12})} + \frac{\tau_{21} \exp(-\alpha\tau_{21})}{x_1 + x_2 \exp(-\alpha\tau_{21})} \right). \tag{2.22}$$

Herein, the binary interaction parameters are denoted by τ_{12} and τ_{21} and the non-randomness parameter by α. The influence of the pressure on the Gibbs free energy of a liquid phase is neglected by the model. As the NRTL model of excess Gibbs free energy G^E can be nonconvex, it allows to model LLE which at the same time represents a risk because it may result in undesired nonconvexities. As a result, parameter estimation for G^E models that allow for nonconvexities guaranteeing thermodynamic stability poses a challenging task, see e.g. Chapter 3.

In BOARPET the model equations for the parameters are translated and scaled by $tran$ and T_{sc}, respectively (57; 56; 12). The model for τ_{12} as a function of temperature T in Kelvin reads

$$\tau_{12}(T) = A_{12}^{BOARPET} + B_{12}^{BOARPET}\left(\frac{T_{\text{sc}}}{T} - tran\right) + C_{12}^{BOARPET}\left(\frac{T}{T_{\text{sc}}} - tran\right) + D_{12}^{BOARPET}\ln\left(\frac{T}{T_{\text{sc}}}\right). \quad (2.23)$$

Similarly, the model for α reads

$$\alpha(T) = E_{12}^{BOARPET} + F_{12}^{BOARPET}\left(\frac{T}{T_{\text{sc}}} - tran\right), \quad (2.24)$$

where $tran = 1$ and $T_{\text{sc}} = 298.15\,\text{K}$.

In Aspen Plus (7) the model equations for the parameters read

$$\tau_{12}(T) = a_{12}^{Aspen} + \frac{b_{12}^{Aspen}}{T} + e_{12}^{Aspen}\ln(T) + f_{12}^{Aspen}T, \quad (2.25)$$

$$\alpha(T) = c_{12}^{Aspen} + d_{12}^{Aspen}(T - 273.15\,\text{K}). \quad (2.26)$$

3. Methods for regressing diffusive equilibria

In this chapter, examples for constraints included in the regression of liquid-liquid equilibria are presented. For different types of objective functions employed in parameter estimation, see Appendix A.

3.1. Formulations without strict isoactivity condition

In this section, formulations that do not guarantee satisfaction of the isoactivity condition (2.21) concerning the regressed model at the predicted mole fractions are discussed. One type of these formulations includes the violation of the isoactivity equation as a penalty term into the objective function. In this case, the 1st-order necessary KKT conditions of the flash problem (see Equation 2.11) are only fulfilled, if the isoactivity term in objective function equals zero. Alternatively, isoactivity is imposed as an inequality constraint. Marcilla et al. (51) report violation of the isoactivity condition by the binary interaction parameter (BIP) regression tool in CHEMCAD 5.2.0. (19). Additionally, the DRS regression tool in Aspen Plus does not fulfil the isoactivity condition for all predictions in one of the case studies presented in Chapter 4.

3.2. 1st-order necessary KKT conditions

Most-commonly, the 1st-order KKT conditions of the flash problem (see Equations (2.11)) or the isoactivity condition (see Equation (2.21)), which can be derived from the 1st-order KKT conditions, are imposed as constraints in parameter estimation. However, 1st-order KKT conditions do not guarantee stability. A parameter estimation method with isoactivity conditions implemented as constraints is described by e.g. Wang et al. (91).

3.3. 2nd-order sufficient KKT conditions

The 2nd-order sufficient KKT conditions as stated in Section 2.1 are necessary and sufficient conditions for a strict local minimum of Gibbs free energy. Local optima of Gibbs free energy that are not strict local optima correspond to the limit of thermodynamic stability, i.e., in a binary system the second derivative of Gibbs free energy with respect to the mole fraction of any species in any phase is zero at a stationary point, see e.g. Glass et al. (33). In this case derivatives of higher order are necessary to decide on stability, see e.g. Tester and Modell (86). Additionally, the 2nd-order sufficient KKT conditions for strict local minima are necessary conditions for strict global minima. These conditions translate to strict convexity of the function of Gibbs free energy at the predicted phase compositions, see e.g. Glass et al. (33). For example, Stragevitch (82) presents a tool for the regression of LLE by minimising a Maximum Likelihood objective function subject to the isoactivity condition and the convexity of the surface of Gibbs free energy at the predicted compositions.

3.4. Correct numbers of phase splits and phases per phase split

Some authors suggest using further constraints in addition to 1st-order or 2nd-order KKT conditions, in order to reduce the risk to predict the wrong numbers of phase splits and phases per phase split. In some of the suggested formulations convexity of Gibbs free energy is imposed on a discrete grid over the composition ranges outside the expected phase splits, whereas in others, penalty terms are used to avoid nonconvexity of the flash problems resulting from the obtained parameters. A method that is supposed to check for convexity of Gibbs free energy on a discrete grid, at least to identify a phase split within flash calculations, is available through the Gauss-Newton algorithm in the DECHEMA Data Preparation Package (DPP) (23). This formulation uses proprietary implementations based on routines described by Novak et al. (61) in combination with routines from the prior DECHEMA tool DETHERM-SDC (87). As Novak et al. (61) state, the method described [ibid.] calculates the second derivative of the surface of Gibbs free energy for a number of compositions to check for convexity. If all second derivatives are greater than zero, the studied system is assumed to be homogeneous on the *continuous* set of mole fractions, yet represented by a *discrete* set.

A different approach is presented by Sørensen and Arlt (78). Therein, two terms are added to the Least Squares objective function to improve different aspects of the fit. Firstly, a penalty term is included into the objective function to reduce the risk of nonconvexity. The included penalty term is $q\mathbf{p}^T\mathbf{p}$, where q is an empirically-chosen constant and $\mathbf{p}$ is the parameter vector. Secondly, if a good estimation of the distribution ratio d_i of species i is of interest the following term is included [ibid.] into the objective function for a ternary system with two phases:

$$\sum_{k\in K}\left[\ln\left(d_i\frac{\gamma^{\infty}_{i,1,k}}{\gamma^{\infty}_{i,2,k}}\right)\right]^2 . \tag{3.1}$$

Herein, $\gamma^{\infty}_{i,1,k}$ denotes the predicted activity coefficient of species i at infinite dilution in phase one of experiment k. Furthermore, Sørensen and Arlt (78) check for necessary criteria including the isoactivity condition and the convexity of the surface of Gibbs free energy at the predicted mole fractions.

3.5. Bilevel problem

If the global solution of the flash problem (see Equations (2.9a) - (2.9c)) is incorporated as necessary and sufficient condition for stability of the predictions into a parameter estimation problem for a model without necessity for root discrimination, the result is a bilevel problem that reads as follows, see e.g. Glass et al. (36)

$$\min_{\mathbf{x}^u\mathbf{p}^u} f^u\left(\mathbf{x}^u,\mathbf{p}^u\right) \tag{3.2a}$$

$$\text{s.t. } \mathbf{h}^u\left(\mathbf{x}^u,\mathbf{p}^u\right)=0, \tag{3.2b}$$

$$\left.\begin{aligned} &f^{l,llp}\left(\mathbf{x}^{u,llp}\mathbf{p}^u\right)\geqslant 0\\ &\mathbf{x}^{u,llp}\in \underset{\mathbf{x}^{l,llp}}{\operatorname{argmin}} f^{l,llp}\left(\mathbf{x}^{l,llp},\mathbf{p}^u\right)\\ &\qquad\text{s.t. } \mathbf{g}^{l,llp}\left(\mathbf{x}^u,\mathbf{x}^{l,llp},\mathbf{p}^u\right)\leqslant 0\end{aligned}\right\}\forall llp\in LLP \tag{3.2c}$$

$$\dots$$

$$\mathbf{x}^{l,llp}\in X^{l,llp},\mathbf{x}^{u,llp}\in X^{u,llp},\mathbf{x}^u\in X^u,\mathbf{p}^u\in P^u,$$

where f^u denotes the objective function of the upper-level problem (ulp), $\mathbf{x}^u\in X^u$ denotes the matrix of mole fraction variables of the ulp, $\mathbf{p}^u\in P^u$ denotes the vector of thermodynamic parameters which are also regression variables of the ulp, $\mathbf{h}^u$ denotes the isoactivity condition, $f^{l,llp}$ denotes the objective function of the lower-level problem (llp), $\mathbf{x}^{u,llp} \in X^{u,llp}$ denotes the matrix of upper-level variables of the llp, $\mathbf{x}^{l,llp} \in X^{l,llp}$ denotes the matrix of lower-level variables of the llp, $\mathbf{g}^{l,llp}$ denotes the inequality constraints of the llp and LLP denotes the index set of llp. The set of llp includes problems that ensure the correct numbers of phase splits and phases per phase split and Baker's criterion, based on the works of Mitsos and Bollas (57; 56; 12). If in addition the ulp is solved to global optimality, this results in a globally optimal fit. An example for the implementation of such a method is BOARPET (58).

4. Regression results for TMFB case studies

Besides parameter estimations for the NRTL model also regressions for the Universal Quasi-Chemical (UNIQUAC) model of Abrams and Prausnitz (1) are included in the literature reviews for case studies in the present chapter. An overview of regression results that are reported for the TMFB case studies in the open literature is given in Appendix B.

4.1. Methods and materials for regression in Aspen Properties

We use Aspen Plus V8.8 (7) with the option settings reported in Table 4.1.The default settings for standard deviations σ_k^T and $\sigma_{i,j,k}^x$ of experimental data are used, as fitting in Aspen is conducted to reverse-engineer the regression formulation focusing on the constraints. The author does not intend to gain a particularly good fit and will not recommend these parameter values for use. The option "Test Stability of LLE results" is assumed to activate the check of the DRS results by comparison to the three-phase flash results, which might be a similar solver for flash problems as included in the FLASH3 unit operation model. The bounds and initial guesses used for regression in Aspen Plus in the present work are presented in Table 4.2.

The bounds specified in Table 4.2 for b_{12}^{Aspen} and b_{21}^{Aspen} are defaults in Aspen Plus. Additionally, the parameters d_{12}^{Aspen}, e_{12}^{Aspen} and e_{21}^{Aspen} are fixed to zero for all case studies. There is an option to impose binary parameters to be symmetric. But activating this option for c_{12}^{Aspen} leads to an error. Nevertheless, in the Aspen Help only a_{ij}^{Aspen},b_{ij}^{Aspen},e_{ij}^{Aspen} and f_{ij}^{Aspen} are stated to be unsymmetrical. Hence, c_{ij}^{Aspen} and d_{ij}^{Aspen} are only specified for one combination of indices in the parameter window.

Table 4.1.: Options used for regression in Aspen Plus DRS

Option	Setting
Perform one dimensional search:	enabled
Test Stability of LLE results:	enabled
Use pre-Release 8 algorithm:	disabled
Initialisation algorithm for butanone/water:	Weighted Least Squares
Initialisation algorithm for all other case studies:	none
Main algorithm:	Britt-Luecke
Maximal #iterations initialisation for butanone/water:	6000
Maximal #iterations for main algorithm:	500
Maximal #iterations for middle loop:	20
Maximum parameter adjustment step size:	0.1
Numerical derivative step size:	0.0001
Convergence tolerance:	0.0001
Scale factor for all parameters:	1
Standard deviation of temperatures σ_k^T:	$0.01\,\mathrm{K}$
Standard deviation of mole fractions $\sigma_{i,j,k}^x$:	$0.001 x_{i,j,k}^{\mathrm{exp}}$
Weights for all data sets:	1

Table 4.2.: Bounds and initial guesses for regressed parameters in Aspen Plus DRS

Parameter	Lower bound	Initial guess	Upper bound
a_{12}^{Aspen}	-10000	0	10000
b_{12}^{Aspen}	-10000	0	10000
f_{12}^{Aspen}	-10000	0	10000
c_{12}^{Aspen}	0.1	0.2	0.5
a_{21}^{Aspen}	-10000	0	10000
b_{21}^{Aspen}	-10000	0	10000
f_{21}^{Aspen}	-10000	0	10000

The following error messages are encountered during regression with Aspen Plus:

I: * ERROR:
REGRESSION RESULTS DO NOT AGREE WITH 3-PHASE FLASH RESULTS.
RERUN WITH DIFFERENT PARAMETER INITIAL GUESSES.

II: ** ERROR
DRS RESULTS DO NOT AGREE WITH THREE-PHASE FLASH RESULTS.
- RERUN WITH A DIFFERENT SET OF PARAMETER INITIAL VALUES
- RECHECK THE DATA IN THE DATA-GROUP PARAGRAPHS
- CHECK STD-DEV FOR SPECIFICATIONS THAT ARE TOO LARGE

III: ** ERROR
CONSTRAINTS CANNOT BE SATISFIED FOR THIS CASE.

IV: ***SEVERE ERROR
FORTRAN INVALID OPERATION ENCOUNTERED.

Usually error messages are shown in the status window of the respective regression and in the history file. However, sometimes results are available without errors reported in the status window, but the history file still includes tables with data points for which the DRS results and three-phase flash results do not agree. Nevertheless, the first error listed above is recorded in the report file, if the history file shows tables with data points for which the DRS results and three-phase flash results do not agree. The errors II to IV are shown both in the status window and in the history file. In case the second error is observed, the history file shows tables with data points for which the DRS results and three-phase flash results do not agree. Obviously, errors I and II can be avoided by deactivating the option "Test Stability of LLE results", but this is not recommended by the author. To avoid the third error changing the initialisation algorithm or its maximal number of iterations seems helpful. To avoid the fourth error changing the number of regressed variables is sometimes helpful. Moreover, the following warnings are encountered relatively often in the status window and in the history file:

I: * WARNING
MAXIMUM NUMBER OF ITERATIONS OF 50 EXCEEDED

II: * WARNING LINE SEARCH FAILURE FOR THIS CASE.

III: * WARNING CONSTRAINTS NOT (TIGHTLY) SATISFIED FOR THIS CASE.

In the first warning, 50 is the current maximal number of iterations for the main algorithm. Obviously, this warning can often be avoided by increasing the maximum number of iterations. Warnings II and III can often be avoided by changing the initialisation method. In addition to errors and warnings, crashes are observed regularly, when executing the Aspen Plus DRS regression tool. For all case studies except butanone/water, the intialisation method is turned off by setting the maximal number of iterations to zero as this reduces the error messages and avoids crashes that occur using the initialisation algorithms Deming and Weighted Least Squares for the same parameter estimation problems. The objective function utilised for parameter estimation in Aspen DRS is given in Equation (A.6).

4.2. Methods and materials for regression in BOARPET

To obtain parameter sets for the NRTL model, we use BOARPET v1.4 (58) calling BARON (83) with GAMS Release 24.8.4 (28) running a Windows Server 2008 R2 Enterprise 64-bit operating system on a machine with an Intel Xeon CPU E5-2640 v3 at 2.6 GHz. The BARON options are set to: AbsConFeasTol 1e-007, RelConFeasTol 0, boxtol 1e-005,firstfeas 0, CompIIS 1, PDo -1, NoutIter 16, NoutPerVar 8, NodeSel 1, lpsol 3, nlpsol -1, numloc 10. The objective function utilised for parameter estimation in BOARPET is given in Equation (A.3).

4.3. LLE 2,5-DMF/water

The purpose of this section is to document the complete collection of measurement data that are currently available in the open literature, and to report parameter values of the NRTL model, for the liquid-liquid equilibrium (LLE) of the binary fluid system of 2,5-dimethylfuran and water (2,5-DMF/water). To the author's best knowledge, this work is the first to assemble all measurements. The parameters are generated using the software BOARPET (58) which is based on the work of Mitsos and Bollas (57; 56; 12). For the regressed parameters, the method guarantees satisfaction of Baker's criterion (8), the correct number of phase splits, correct number of phases per phase split and other desired behaviour. The regression results in a good fit. NRTL parameters are reported that allow modelling of the LLE for the temperature range $283.15\,\text{K} \leqslant T \leqslant 313.2\,\text{K}$.

Measurement data for the LLE 2,5-DMF/water are reported by Dechambre (21), Silva and Aznar (20) and Eidinoff (25). Note that since the normal boiling point of 2,5-DMF is reached at around $367\,\text{K}$, sophisticated measurement techniques accounting for elevated pressure levels are required above this temperature.

Da Silva et al. (20) report a phase equilibrium model for this fluid system. They fit NRTL and UNIQUAC models to ternary LLE data that are presented in the same publication. The temperatures for the LLE measurements are $298.2\,\text{K}$ and $313.2\,\text{K}$. The authors conduct three regressions for each model, one of which with respect to LLE data including 2,5-DMF and water in combination with one of the following alcohols: 1-butanol, 2-butanol, 1-hexanol.

The binary interaction parameter of the NRTL model is calculated as $\tau_{ij} = \dfrac{b_{ij}}{T}$ and the only regressed parameters of the NRTL model b_{12} and b_{21}. The energy interaction parameter τ_{ij} of the UNIQUAC model is given by the equation $\tau_{ij} = \exp\left(-\dfrac{a_{ij}}{T}\right)$ and the only regressed parameters in the UNIQUAC model are a_{12} and a_{21}.

Da Silva et al. (20) conduct the regressions with the FORTRAN code TML-LLE 2.0. The reference given for this code is the thesis of Stragevitch (82) (in Portuguese), but details on regression method and problem formulation are not provided, with the exception of the information that the authors apply the Least Squares objective function and solve the optimisation problem with a modified Simplex method. In the Chapters 2.1 and 2.2 in the thesis of Stragevitch (82) the Equations that establish the employed Maximum Likelihood regression model for LLE are stated. In Appendix A.5 the applied local stability test is explained (82). Here the isoactivity condition and the convexity of the surface of Gibbs free energy are listed as necessary criteria. In the thesis of Stragevitch (82) the parameters are regressed by applying a Maximum Likelihood function subject to isoactivity condition and local convexity of Gibbs free energy for the predicted compositions. The problem is solved with a modified Simplex method that is published by Nelder and Mead (59).

Qureshi et al. (64) report activity coefficients for 2,5-DMF at infinite dilution in water for the temperature interval $273.35\,\text{K} \leqslant T \leqslant 333.15\,\text{K}$.

In contrast to Stragevitch (82), the present work reports parameters that are regressed guaranteeing thermodynamic stability by means of a bilevel problem formulation, i.e., a parameter fitting problem subject to criteria that are both sufficient and necessary for stability. This includes Baker's criterion (8), as well as criteria to guarantee the correct number of phases and phase splits. As opposed to the work of Stragevitch (82), the parameter fitting is conducted with the global solver BARON (83) and thus, a globally optimal fit is obtained upon convergence.

4.3.1. Measurement data

The measurement data of Dechambre (21), Silva and Aznar (20) and Eidinoff (25) are reported in Table D.1 and used for regression of the LLE 2,5-DMF/water. Note that, in the present work, no distinction is made between the standard uncertainties and standard deviations reported in the literature and thus, both are denoted by u. The standard uncertainties for the temperature are $u(T) = 0.01\,\mathrm{K}$ (21) and $u(T) = 0.1\,\mathrm{K}$ (20). Eidinoff (25) reports $0.05\,\mathrm{K}$ as appr. uncertainty of the temperature. Please find further information about the treatment of uncertainties and the accuracy of values reported in the present work in Appendix C.

4.3.2. Parameter values

In this subsection, we report the regressed parameter values. Together with the NRTL model from Renon and Prausnitz (65) and the temperature model for the parameters as given by Equations (2.23) and (2.24), this allows modelling of phase equilibria. The indices are assigned as follows: 2,5-DMF (1), H_2O (2).

$A_{12}^{BOARPET} = 3.4358$	*regressed*
$B_{12}^{BOARPET} = 6.9714$	*regressed*
$C_{12}^{BOARPET} = 0.0000$	*fixed*
$D_{12}^{BOARPET} = 0.0000$	*fixed*
$E_{12}^{BOARPET} = 0.2000$	*fixed*
$F_{12}^{BOARPET} = 0.0000$	*fixed*
$A_{21}^{BOARPET} = 5.3850$	*regressed*
$B_{21}^{BOARPET} = -13.715$	*regressed*
$C_{21}^{BOARPET} = 0.0000$	*fixed*
$D_{21}^{BOARPET} = 0.0000$	*fixed*

- Range of validity

$$283.15\,\mathrm{K} \leqslant T \leqslant 313.2\,\mathrm{K}$$

- Objective function value: 0.000017434
- Aqueous phase: $w_{j,k} = 10$
- Organic phase: $w_{j,k} = 1$

4.3.3. Predictions and goodness of fit

For the predicted mole fractions, see Table E.1. Note that the temperature is kept constant for regression. The regression results in a good fit, see top diagram in Figure 4.1. Additionally, parts of the mole fraction range shown in this diagram are presented enlarged in the bottom diagrams in Figure 4.1.

4.4. LLE 2,5-DMTHF/water

The purpose of this section is to document the complete collection of measurement data that are currently available in the open literature, and to report parameter values of the NRTL

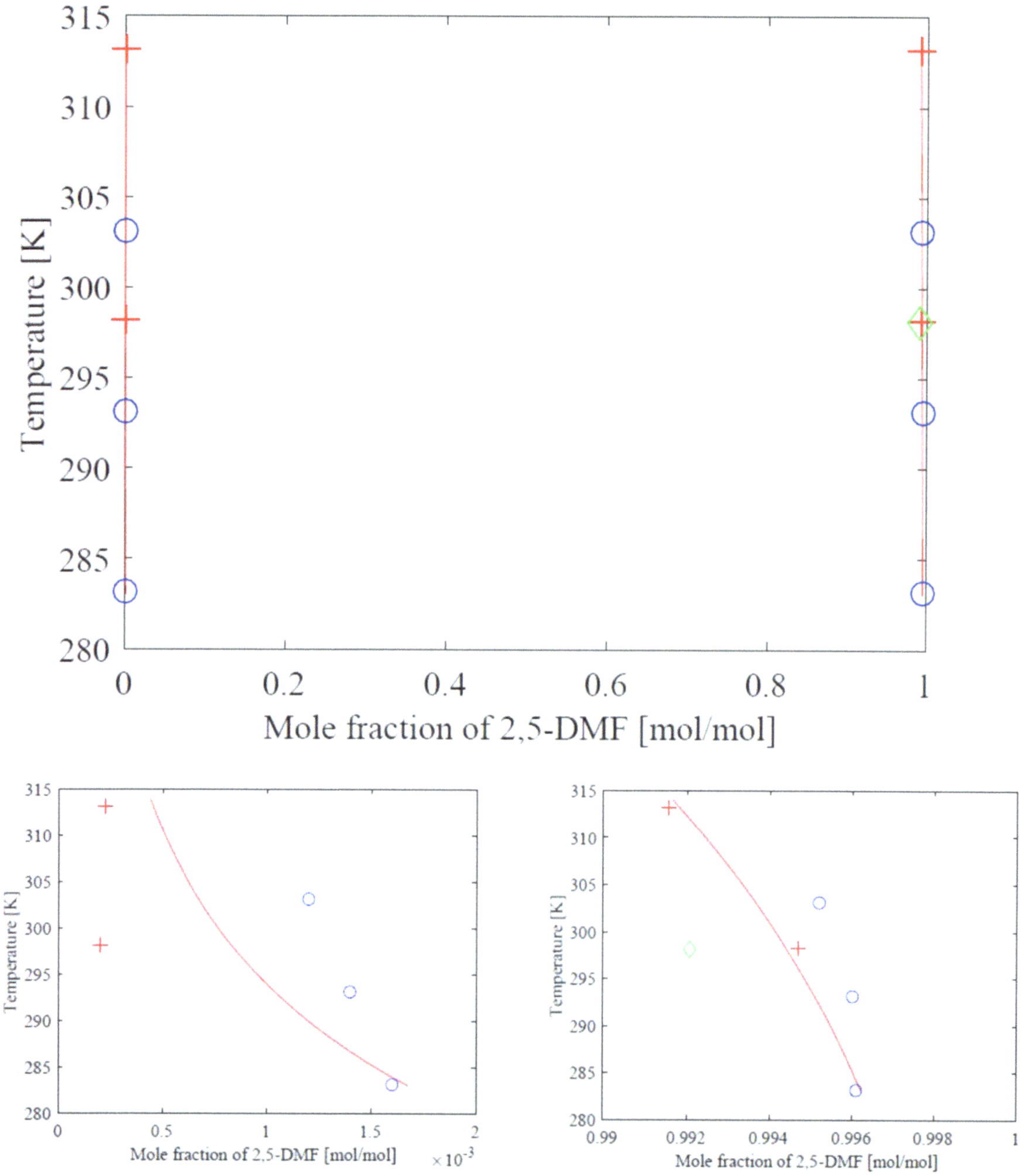

Fig. 4.1.: T-x diagrams of the LLE 2,5-DMF/water. Data are taken from Dechambre (21) (blue circle), Silva and Aznar (20) (red plus) and Eidinoff (25) (green diamond). The mole fractions predicted with the parameters from Subsection 4.3.2 are plotted as red lines. The system 2,5-DMF/water has a broad miscibility gap over the whole temperature range, $283.15\,\mathrm{K} \leqslant T \leqslant 313.2\,\mathrm{K}$, and the regression results in a good fit.

model, for the liquid-liquid equilibrium of the binary fluid system of 2,5-dimethyltetrahydrofuran and water (2,5-DMTHF/water). The substance regarded here as 2,5-DMTHF is a mixture of cis-2,5-dimethyltetrahydrofuran and trans-2,5-dimethyltetrahydrofuran. To the author's best knowledge, this work is the first $i)$ to report a model for this phase equilibrium, and $ii)$ to assemble all measurement data. The parameters are generated using the software BOARPET (58) which is based on the work of Mitsos and Bollas (57; 56; 12). For the regressed parameters, the method guarantees satisfaction of Baker's criterion (8), the correct number of phase splits, correct number of phases per phase split and other desired behaviour. The regression results in a good fit. NRTL parameters are reported that allow modelling of the LLE for the temperature range $273.15\,\mathrm{K} \leqslant T \leqslant 356.15\,\mathrm{K}$.

Measurement data for the LLE 2,5-DMTHF/water are reported by Dechambre (21) and Stephenson (81). Note that since the normal boiling point of 2,5-DMTHF is reached at around $364\,\mathrm{K}$, sophisticated measurement techniques accounting for elevated pressure levels are required above this temperature. Cabani et al. (17) report the activity coefficient for 2,5-DMTHF at infinite dilution in water at $298.15\,\mathrm{K}$ and $101\,325\,\mathrm{Pa}$.

The present work reports parameters that are regressed guaranteeing thermodynamic stability by means of a bilevel problem formulation, i.e., a parameter fitting problem subject to criteria that are both sufficient and necessary for stability. This includes Baker's criterion (8), as well as criteria to guarantee the correct number of phases and phase splits. The parameter fitting is solved with the global solver BARON (83) and thus, a globally optimal fit is obtained upon convergence.

4.4.1. Measurement data

The measurement data of Dechambre (21) and Stephenson (81) are reported in Table D.2 and used for regression of the LLE 2,5-DMTHF/water. Note that, in the present report, no distinction is made between the standard uncertainties and standard deviations reported in the literature and thus, both are denoted by u. The standard uncertainty for the temperature is given as $u(T) = 0.01\,\mathrm{K}$ by Dechambre (21). Stephenson (81) reports $0.1\,\mathrm{K}$ as appr. uncertainty of the temperature. Please find further information about the treatment of uncertainties and the accuracy of values reported in the present work in Appendix C.

4.4.2. Parameter values

In this subsection, we report the regressed parameter values. Together with the NRTL model from Renon and Prausnitz (65) and the temperature model for the parameters as given by Equations (2.23) and (2.24), this allows modelling of phase equilibria. The indices are assigned as follows: 2,5-DMTHF (1), H_2O (2).

$$A_{12}^{BOARPET} = 1.5043 \qquad regressed$$
$$B_{12}^{BOARPET} = 1.3754 \qquad regressed$$
$$C_{12}^{BOARPET} = 0.0000 \qquad fixed$$
$$D_{12}^{BOARPET} = 0.0000 \qquad fixed$$
$$E_{12}^{BOARPET} = 0.28459 \qquad regressed$$
$$F_{12}^{BOARPET} = 0.0000 \qquad fixed$$
$$A_{21}^{BOARPET} = 3.8403 \qquad regressed$$
$$B_{21}^{BOARPET} = -7.9673 \qquad regressed$$
$$C_{21}^{BOARPET} = 0.0000 \qquad fixed$$
$$D_{21}^{BOARPET} = 0.0000 \qquad fixed$$

- Range of validity

$$273.15\,\text{K} \leqslant T \leqslant 356.15\,\text{K}$$

- Objective function value: 0.0010269
- Aqueous phase: $w_{j,k} = 10$
- Organic phase: $w_{j,k} = 1$

4.4.3. Predictions and goodness of fit

For the predicted mole fractions, see Table E.2. Note that the temperature is kept constant for regression. The regression results in a good fit, see top diagram in Figure 4.2. Additionally, the lower part of the mole fraction range shown in this diagram is presented enlarged in the bottom diagram in Figure 4.2.

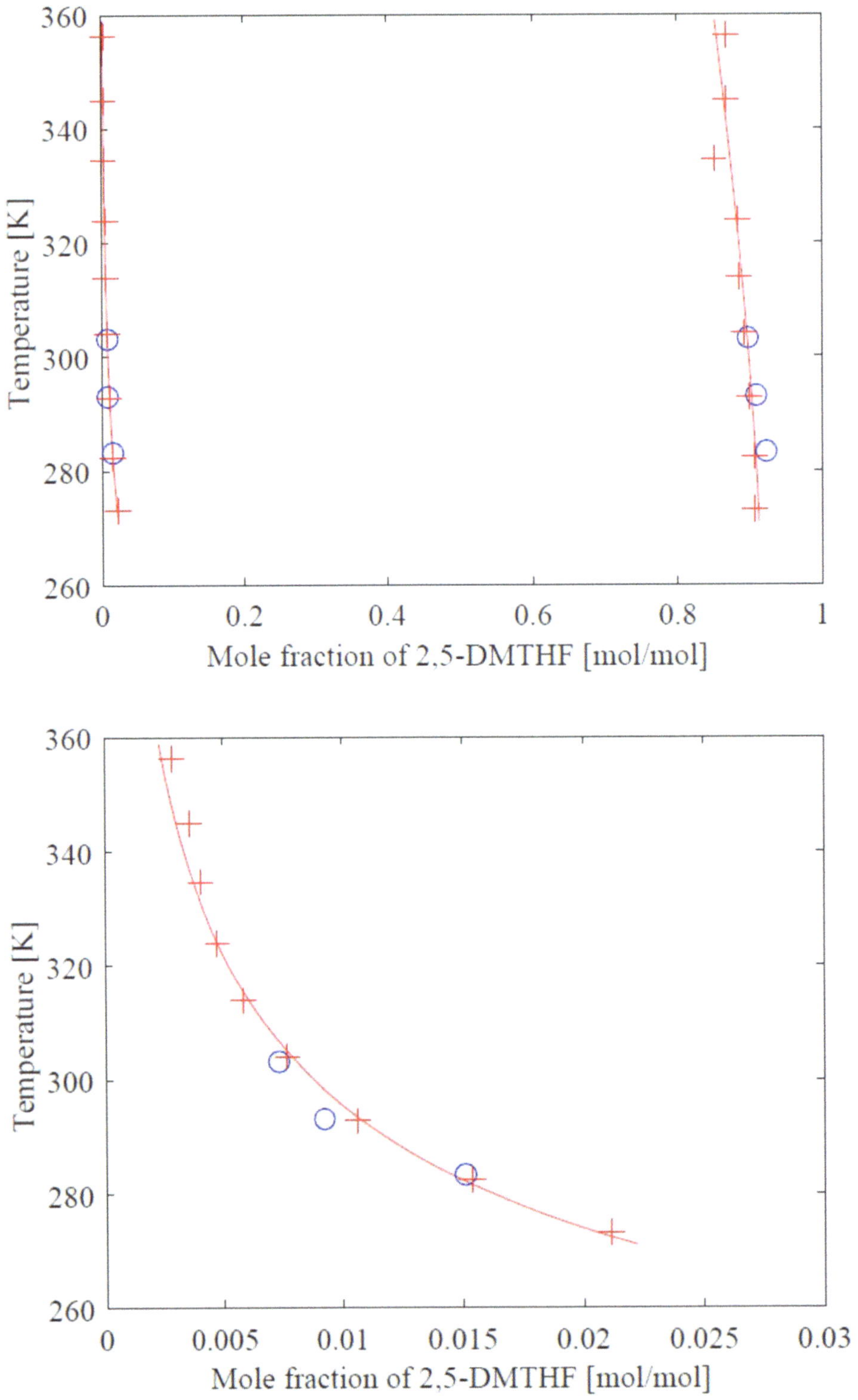

Fig. 4.2.: T-x diagrams of the LLE 2,5-DMTHF/water. Data are taken from Dechambre (21) (blue circle) and Stephenson (81) (red plus). The mole fractions predicted with the parameters from Subsection 4.4.2 are plotted as red lines. The regression results in a good fit.

4.5. LLE 2-MF/water

The purpose of this section is to document the complete collection of measurement data that are currently available in the open literature, and to report parameter values of the NRTL model, for the liquid-liquid equilibrium (LLE) of the binary fluid system of 2-methylfuran and water (2-MF/water). To the author's best knowledge, this work is the first to assemble all measurements. The parameters are generated using the software BOARPET (58) which is based on the work of Mitsos and Bollas (57; 56; 12). For the regressed parameters, the method guarantees satisfaction of Baker's criterion (8), the correct number of phase splits, correct number of phases per phase split and other desired behaviour. The regression results in a good fit. NRTL parameters are reported that, under some assumptions, allow modelling of the LLE for the temperature range $283.15\,\text{K} \leqslant T \leqslant 328.15\,\text{K}$.

Measurement data for the LLE 2-MF/water are reported by Dechambre (21), Smith and LaBonte (73), Uusi-Kyyny et al. (88) and Karmilchik and Efimova (43). Note that since the normal boiling point of 2-MF is reached at around $338\,\text{K}$, sophisticated measurement techniques accounting for elevated pressure levels are required above this temperature. Furthermore, Uusi-Kyyny et al. (88) report that decomposition products of 2-MF are detected in the organic phase of this LLE system. The assumption of a binary system is used [ibid.], since the decomposition products are not detected in the aqueous phase.

Uusi-Kyyny et al. (88) fit NRTL and UNIQUAC models simultaneously to LLE and vapour-liquid-liquid equilibrium (VLLE) data of this system. For both models two regressions with four and six regressed parameters each are conducted. The experimental data that are fitted in their regression are binary VLLE data of Smith and LaBonte (73) and composition data of the aqueous phase in a binary LLE measured by Uusi-Kyyny et al. (88). The regression is conducted in Aspen Plus V8.6 (6) and for the fugacities required for VLLE models the Peng Robinson equation of state is employed. Neither the objective function nor the constraints employed in the regression are mentioned. The temperature boundaries, within which the model parameters are valid, are given by Uusi-Kyyny et al. (88) as $298\,\text{K} \leqslant T \leqslant 415\,\text{K}$.

Sørensen and Arlt (77) fit NRTL and UNIQUAC models to LLE data of the ternary systems 2-MF/water/tetrahydrofuran and 2-MF/water/2-furylmethanol that are reported by Boldyrev et al. (11). As reported in Sørensen and Arlt (78), they only check for necessary (as opposed to sufficient) criteria for thermodynamic stability. These criteria include the isoactivity equations and the convexity of the surface of Gibbs free energy at the predicted mole fractions. Moreover, a penalty term is included into the objective function to reduce the risk of nonconvexity according to the authors.

Aneja et al. (3) calculate the solubility of 2-MF in water based on octanol/water partition coefficients. Furthermore, Uusi-Kyyny et al. (88) compute solubilities from Henry's law constants given by Wieland et al. (93) for the binary system 2-MF/water. Qureshi et al. (64) report activity coefficients for 2-MF at infinite dilution in water for the temperature interval $273.35\,\text{K} \leqslant T \leqslant 333.15\,\text{K}$.

In contrast to the methods referenced in this subsection, the present work reports parameters that guarantee thermodynamic stability by means of a bilevel problem formulation, i.e., a parameter fitting problem subject to criteria that are both sufficient and necessary for stability. This includes Baker's criterion (8), as well as criteria to guarantee the correct number of phases and phase splits. The parameter fitting is conducted with the global solver BARON (83) and thus, a globally optimal fit is obtained upon convergence.

4.5.1. Measurement data

The measurement data of Dechambre (21), Smith and LaBonte (73), Uusi-Kyyny et al. (88) and Karmilchik and Efimova (43) are reported in Table D.3 and used for regression of the LLE 2-MF/water. The solubility of 2-MF in water that is calculated by Aneja et al. (3) and the solubilities that are computed by Uusi-Kyyny et al. (88) from Henry's law constants are excluded from the regression because these data are not measured in an LLE of the 2-MF/water system. Note that, in the present work, no distinction is made between the standard uncertainties and standard deviations reported in the literature and thus, both are denoted by u. The standard uncertainty for the temperature is given by $u(T) = 0.01\,\text{K}$ by Dechambre (21). Uusi-Kyyny et al. (88) report $0.02\,\text{K}$ as appr. uncertainty of the temperature. Please find further information about the treatment of uncertainties and the accuracy of values reported in the present work in Appendix C.

4.5.2. Parameter values

In this subsection, we report the regressed parameter values. Together with the NRTL model from Renon and Prausnitz (65) and the temperature model for the parameters as given by Equations (2.23) and (2.24), this allows modelling of phase equilibria. The indices are assigned as follows: 2-MF (1), H_2O (2). Due to difficulties of the regression of incomplete measurement data that are specified in Subsection 4.5.1, two sets of parameters are reported here.

Run 1

$A_{12}^{BOARPET} = 1.8821$ *regressed*

$B_{12}^{BOARPET} = 22.500$ *regressed*

$C_{12}^{BOARPET} = 0.0000$ *fixed*

$D_{12}^{BOARPET} = 0.0000$ *fixed*

$E_{12}^{BOARPET} = 0.10506$ *regressed*

$F_{12}^{BOARPET} = 0.0000$ *fixed*

$A_{21}^{BOARPET} = 4.7127$ *regressed*

$B_{21}^{BOARPET} = -25.859$ *regressed*

$C_{21}^{BOARPET} = 0.0000$ *fixed*

$D_{21}^{BOARPET} = 0.0000$ *fixed*

Run 2

$A_{12}^{BOARPET} = 2.6680$ *regressed*

$B_{12}^{BOARPET} = 7.8039$ *regressed*

$C_{12}^{BOARPET} = 0.0000$ *fixed*

$D_{12}^{BOARPET} = 0.0000$ *fixed*

$E_{12}^{BOARPET} = 0.20908$ *regressed*

$F_{12}^{BOARPET} = 0.0000$ *fixed*

$A_{21}^{BOARPET} = 4.7738$ *regressed*

$B_{21}^{BOARPET} = -4.4167$ *regressed*

$C_{21}^{BOARPET} = 0.0000$ *fixed*

$D_{21}^{BOARPET} = 0.0000$ *fixed*

- Range of validity for both parameter sets

$$283.15\,\mathrm{K} \leqslant T \leqslant 328.15\,\mathrm{K}$$

- Objective function value of Run 1: 0.00018349
- Objective function value of Run 2: 0.00085904
- Aqueous phase: $w_{j,k} = 10$
- Organic phase: $w_{j,k} = 1$

4.5.3. Predictions and goodness of fit

The aim of the regressions using BOARPET (58) is to obtain models that guarantee thermodynamic stability and represent a good fit to the available measurement data. We observed that incomplete measurement data lead to especially challenging parameter estimation problems. The reason is that there are mole fraction variables for the equilibrium conditions which are not part of the objective function. As a consequence of incomplete measurements, the experimental shape of the miscibility gap is unknown and predictions by the model resulting from regression can, in some sense, be considered to represent extrapolations. In order to demonstrate this difficulty, two different sets of parameters are reported in this section but none is preferable over the other. These parameter sets result from two runs with different settings of BOARPET options. Note that it is an open issue whether any of the sets is representative since a comparison to measurement data is not fully applicable. For the results of Run 1 and 2, see Table E.3 and E.4, respectively. Note that the temperature is kept constant for regression. For the measurements and predictions of mole fractions over the whole temperature range of the literature data, see top diagram in Figure 4.3. Additionally, the lower part of the mole fraction range shown this diagram is presented enlarged in the bottom diagram in Figure 4.3.

4.6. LLE 2-MTHF/water

The purpose of this section is to document the complete collection of measurement data that are currently available in the open literature, and to report parameter values of the NRTL

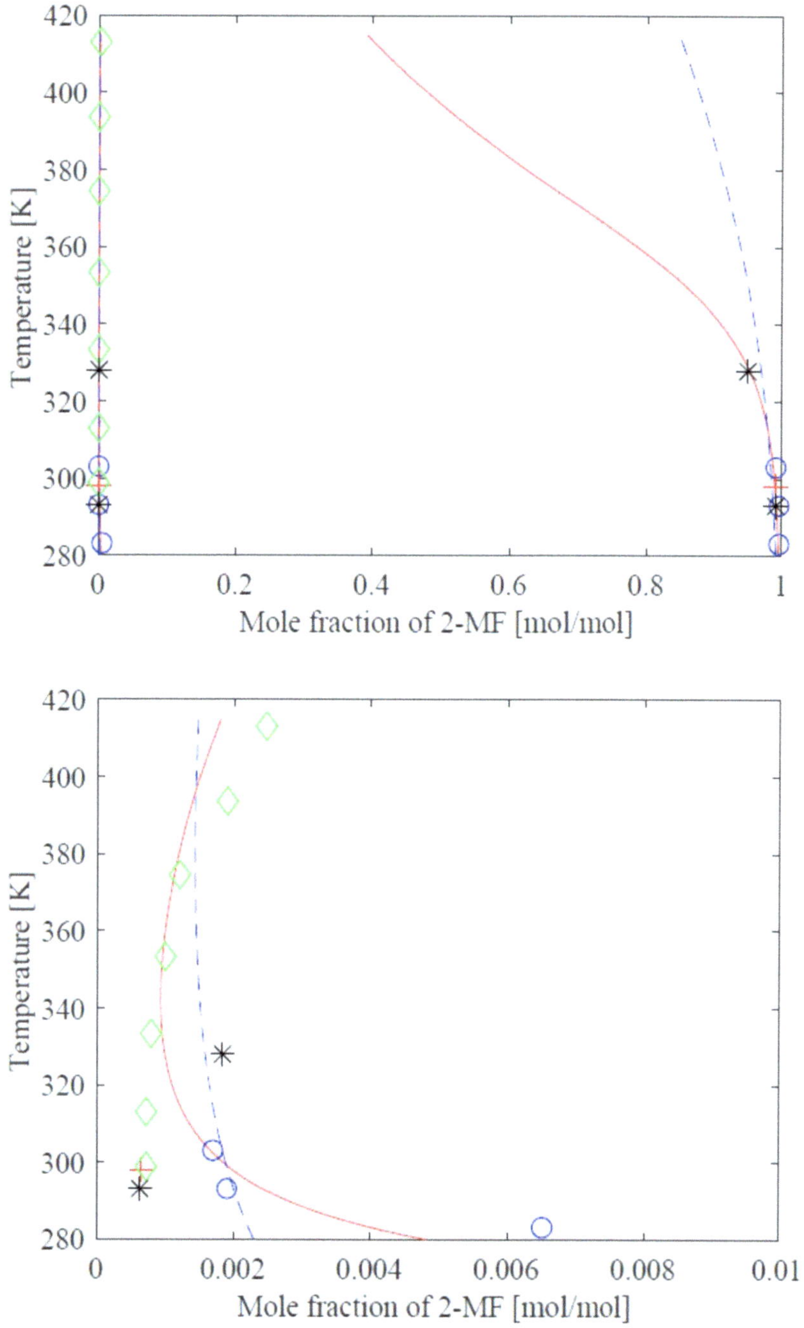

Fig. 4.3.: T-x diagrams of the LLE 2-MF/water. Data are taken from Dechambre (21) (blue circle), Smith and LaBonte (73) (red plus), Uusi-Kyyny et al. (88) (green diamond) and Karmilchik and Efimova (43) (black asterisk). The mole fractions predicted with the parameters from Run 1 are plotted as solid red lines. The mole fractions predicted with the parameters from Run 2 are plotted as dashed blue lines. The system 2-MF/water has a broad miscibility gap between 283.15 K and 328.15 K. Above that temperature interval the measurement data are insufficient to obtain a reliable prediction of the miscibility gap.

model, for the liquid-liquid equilibrium (LLE) of the binary fluid system of 2-methyltetrahydrofuran and water (2-MTHF/water). To the author's best knowledge, this work is the first to assemble all measurements. The parameters are generated using the software BOARPET (58) which is based on the work of Mitsos and Bollas (57; 56; 12). For the regressed parameters, the method guarantees satisfaction of Baker's criterion (8), the correct number of phase splits, correct number of phases per phase split and other desired behaviour. The regression results in a good fit. NRTL parameters are reported that allow modelling of the LLE for the temperature range $273.15\,\text{K} \leqslant T \leqslant 423.15\,\text{K}$.

Measurement data for the LLE 2-MTHF/water are reported by Dechambre (21), Glass et al. (33), Männistö et al. (50), Stephenson (81), Houben et al. (42) and Bennett and Philip (9). Note that since the normal boiling point of 2-MTHF is reached at around $351\,\text{K}$, sophisticated measurement techniques accounting for elevated pressure levels are required above this temperature.

Sørensen and Arlt (76) fit NRTL and UNIQUAC models to smoothed binary LLE data of Bennett and Philip (9) at constant temperatures. Sørensen and Arlt note that they only check for necessary (as opposed to sufficient) criteria of thermodynamic stability. These criteria include the isoactivity equations and the convexity of the surface of Gibbs free energy at the predicted mole fractions.

Männistö et al. (50) regress binary parameters for the UNIQUAC model to data of this binary system and of further binary systems including the substances 2-MTHF, water, furfural and cyclopentyl methyl ether. The predictions of the resulting UNIQUAC model are compared to data of ternary LLE. The regression is conducted with Aspen Plus using the Britt-Luecke algorithm and the Maximum Likelihood objective function. The parameters $A_{ij}, A_{ji}, B_{ij}, B_{ji}, D_{ij}, D_{ji}$ of the employed UNIQUAC-HOC model are regressed, while the parameters $C_{ij}, C_{ji}, E_{ij}, E_{ji}$ are excluded from the regression. The measurement data that are used for the regression by Männistö et al. (50) covers a temperature range from $287.2\,\text{K}$ to $343.2\,\text{K}$. Glass et al. (33) fit NRTL parameters to their own measurements by means of the same method as in the present report. The measurements conducted by Glass et al. cover a particularly broad temperature range, $293.15\,\text{K}$ to $423.15\,\text{K}$. Pressure is assumed to have a negligible influence on the mole fractions in LLE. Finally, Cabani et al. (17) report the activity coefficient for 2-MTHF at infinite dilution in water at $298.15\,\text{K}$ and $101\,325\,\text{Pa}$.

In contrast to Männistö et al. (50) and Sørensen and Arlt (76), the present work reports parameters that are regressed guaranteeing thermodynamic stability by means of a bilevel problem formulation, i.e., a parameter fitting problem subject to criteria that are both sufficient and necessary for stability. This formulation is also applied by Glass et al. (33) to the system 2-MTHF/water and it includes Baker's criterion (8), as well as criteria to guarantee the correct number of phases and phase splits. The parameter fitting is conducted with the global solver BARON (83) and thus, a globally optimal fit is obtained upon convergence.

4.6.1. Measurement data

The measurement data of Dechambre (21), Glass et al. (33), Männistö et al. (50), Stephenson (81) and Houben et al. (42) are reported in Table D.4 and used for regression of the LLE 2-MTHF/water. The LLE data given by Bennett and Philip (9) are excluded from the regression because of inconsistency with the other literature data. Note that, in the present work, no distinction is made between the standard uncertainties and standard deviations reported in the literature and thus, both are denoted by u. The standard uncertainties for the temperature are $u(T) = 0.01\,\text{K}$ (21), $u(T) = 0.3\,\text{K}$ (33) and $u(T) = 0.2\,\text{K}$ (50). Stephenson (81) report $0.1\,\text{K}$ as appr. uncertainty of the temperature. Houben et al. (42) do not report uncertainties. However, they execute redundant measurements of compositions at the same temperatures. The error bars of these measurements are depicted in Figure 4.4, but are hardly visible due to the relatively small variations. Please find further information about the treatment of uncertainties and the accuracy of values reported in the present work in Appendix C.

4.6.2. Parameter values

In this subsection, we report the regressed parameter values. Together with the NRTL model from Renon and Prausnitz (65) and the temperature model for the parameters as given by Equations (2.23) and (2.24), this allows modelling of phase equilibria. The indices are assigned as follows: 2-MTHF (1), H_2O (2).

$$
\begin{aligned}
A_{12}^{BOARPET} &= 0.53233 && \textit{regressed} \\
B_{12}^{BOARPET} &= 2.1144 && \textit{regressed} \\
C_{12}^{BOARPET} &= 0.0000 && \textit{fixed} \\
D_{12}^{BOARPET} &= 0.0000 && \textit{fixed} \\
E_{12}^{BOARPET} &= 0.19191 && \textit{regressed} \\
F_{12}^{BOARPET} &= 0.0000 && \textit{fixed} \\
A_{21}^{BOARPET} &= 3.3351 && \textit{regressed} \\
B_{21}^{BOARPET} &= -6.5188 && \textit{regressed} \\
C_{21}^{BOARPET} &= 0.0000 && \textit{fixed} \\
D_{21}^{BOARPET} &= 0.0000 && \textit{fixed}
\end{aligned}
$$

- Range of validity

$$273.15\,\mathrm{K} \leqslant T \leqslant 423.15\,\mathrm{K}$$

- Objective function value: 0.029597
- Aqueous phase: $w_{j,k} = 10$
- Organic phase: $w_{j,k} = 1$

4.6.3. Predictions and goodness of fit

For the predicted mole fractions, see Table E.5. Note that the temperature is kept constant for regression. The regression results in a good fit, see Fig. 4.4.

4.7. LLE butanone/water

The purpose of this section is to document a representative selection of the measurement data that are currently available in the open literature, and to report parameter values of the NRTL model, for the liquid-liquid equilibrium (LLE) of the binary fluid system butanone/water. The LLE of this system is discussed extensively in the literature including many more measurements

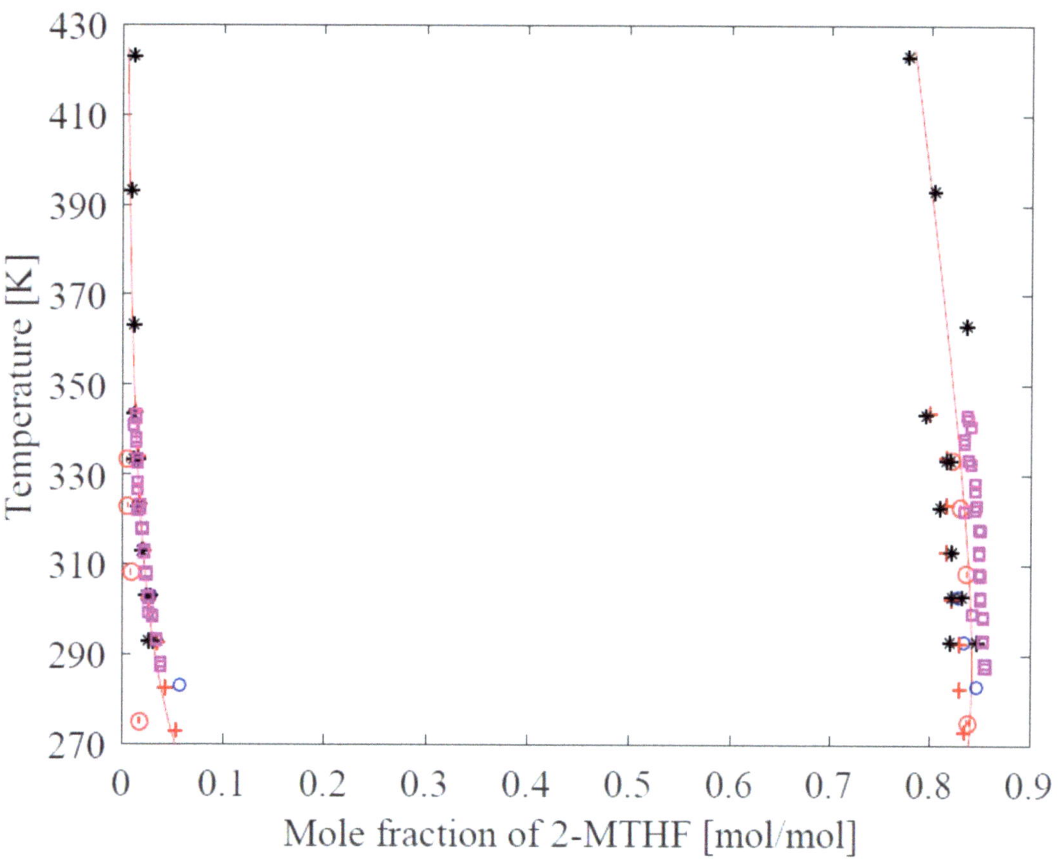

Fig. 4.4.: T-x diagram of the LLE 2-MTHF/water. Data are taken from Dechambre (21) (blue circle), Stephenson (81) (red plus), Houben et al. (42) (red circle), Glass et al. (33) (black asterisk) and Männistö et al. (50) (magenta square). The mole fractions predicted with the parameters from Subsection 4.6.2 are plotted as red lines. The regression results in a good fit.

and models than reported here. The parameters are generated using the software BOARPET (58) which is based on the work of Mitsos and Bollas (57; 56; 12). For the regressed parameters, the method guarantees satisfaction of Baker's criterion (8), the correct number of phase splits, correct number of phases per phase split and other desired behaviour. The regression results in a good fit. NRTL parameters are reported that allow modelling of the LLE for the temperature range $261.75\,\mathrm{K} \leqslant T \leqslant 415.75\,\mathrm{K}$.

Measurement data for the LLE butanone/water are reported by Dechambre et al. (22), Arce et al. (4), Stephenson (81), Ghildiyal et al. (30), Ruiz et al. (69), Ferino et al. (27), Campbell et al. (18), Skrzec and Murphy (72) and many other researchers. Note that since the normal boiling point of butanone is reached at around $353\,\mathrm{K}$, sophisticated measurement techniques accounting for elevated pressure levels are required above this temperature.

Sørensen and Arlt (76) fit NRTL and UNIQUAC models to smoothed literature data of the binary LLE butanone/water at constant temperatures. They point out that they only check for necessary (as opposed to sufficient) criteria for thermodynamic stability. These criteria include the isoactivity equations and the convexity of the surface of Gibbs free energy at the predicted mole fractions. Arce et al. (4) fit NRTL and UNIQUAC models to data of the ternary LLE butanone/water/propanoic acid. In both cases they use a regression tool of Sørensen et al. (79) and a penalty term in the objective function to reduce the risk of multiple local optima. Sørensen et al. (79) argue that, while the global minimum of Gibbs free energy is the necessary and sufficient criterion for stability of LLE at constant temperature and pressure, global optimisation is computationally too expensive and thus, only mass balances, isoactivity equations and convexity of the surface of Gibbs free energy at the predicted compositions are implemented as constraints in their regression tool. Wang et al. (91) report regression results of binary and ternary parameters of the modified UNIQUAC model and the extended UNIQUAC model using LLE data of the binary system butanone/water and of the ternary system butanone/water/diethylcarbonate. The only constraints mentioned [ibid.] are the isoactivity equations and the mass balances. Ochi et al. (62) report regressions of binary LLE including the system butanone/water using the NRTL model. In addition to the isoactivity conditions, the authors discuss further conditions of thermodynamic stability that have to hold at the critical point. These are concerned with the higher-order derivatives of the Gibbs free energy and the enthalpy, respectively, with respect to composition. Ferino et al. (27) report simultaneous correlation of LLE, VLE and excess enthalpy data for binary systems of ketones with water including the system butanone/water using the NRTL model, but do neither give the objective function nor the constraints of the regression. Furthermore, Lladosa et al. (46)

and Ghildiyal et al. (30) correlate LLE data of ternary systems including the species butanone and water using NRTL and UNIQUAC models.

Altogether, in literature, parameter regression is conducted subject to necessary but not sufficient criteria for thermodynamic stability, except for the critical point (62). Hence, stability of the predicted phase equilibrium is not guaranteed. In contrast, the present work reports parameters that are regressed by means of a bilevel problem formulation, i.e., a parameter fitting problem subject to criteria that are both sufficient and necessary for stability of subcritical points. This includes Baker's criterion (8), as well as criteria to guarantee the correct number of phases and phase splits. Thus, thermodynamic stability is guaranteed for all predicted points with the exception of critical points. The parameter fitting is conducted with the global solver BARON (83) and thus, a globally optimal fit is obtained upon convergence.

4.7.1. Measurement data

The measurement data of Dechambre et al. (22), Arce et al. (4), Stephenson (81), Ghildiyal et al. (30), Ruiz et al. (69), Ferino et al. (27), Campbell et al. (18) and Skrzec and Murphy (72) are reported in Table D.5 and used for regression of the LLE butanone/water. Due to the extensive data available for this LLE system, a regression of the complete data available in the open literature is beyond the scope of the present work. The author's selection contains data of the full temperature range in which the LLE of the butanone/water system exists. Only data that are consistent with the majority of available literature data are included.
In the present work, no distinction is made between the standard uncertainties and standard deviations reported in the literature and thus, both are denoted by u. Arce et al. (4), Stephenson (81) and Ghildiyal et al. (30) report 0.1 K as appr. uncertainty of the temperature. Ruiz et al. (69) report 0.05 K as appr. uncertainty of the temperature. Campbell et al. (18) report appr. uncertainties of the temperature between 0.2 K and 0.3 K. Skrzec and Murphy (72) report 0.5 K as appr. uncertainty of the temperature. Please find further information about the treatment of uncertainties and the accuracy of values reported in the present work in Appendix C.

4.7.2. Parameter values

In this subsection, we report the regressed parameter values. Together with the NRTL model from Renon and Prausnitz (65) and the temperature model for the parameters as given by Equations (2.23) and (2.24), this allows modelling of phase equilibria. The indices are assigned as follows: Butanone (1), H_2O (2).

$$
\begin{aligned}
A_{12}^{BOARPET} &= 2.0982 && regressed\\
B_{12}^{BOARPET} &= -14.880 && regressed\\
C_{12}^{BOARPET} &= -15.714 && regressed\\
D_{12}^{BOARPET} &= 0.0000 && fixed\\
E_{12}^{BOARPET} &= 0.42827 && regressed\\
F_{12}^{BOARPET} &= 0.0000 && fixed\\
A_{21}^{BOARPET} &= 2.6072 && regressed\\
B_{21}^{BOARPET} &= -5.8754 && regressed\\
C_{21}^{BOARPET} &= -3.6958 && regressed\\
D_{21}^{BOARPET} &= 0.0000 && fixed
\end{aligned}
$$

- Range of validity

$$261.75\,\mathrm{K} \leqslant T \leqslant 415.75\,\mathrm{K}$$

- Objective function value: 0.046518
- Aqueous phase: $w_{j,k} = 10$
- Organic phase: $w_{j,k} = 1$

4.7.3. Predictions and goodness of fit

For the predicted mole fractions, see Table E.6. Note that the temperature is kept constant for regression. The regression results in a good fit, see Fig. 4.5.

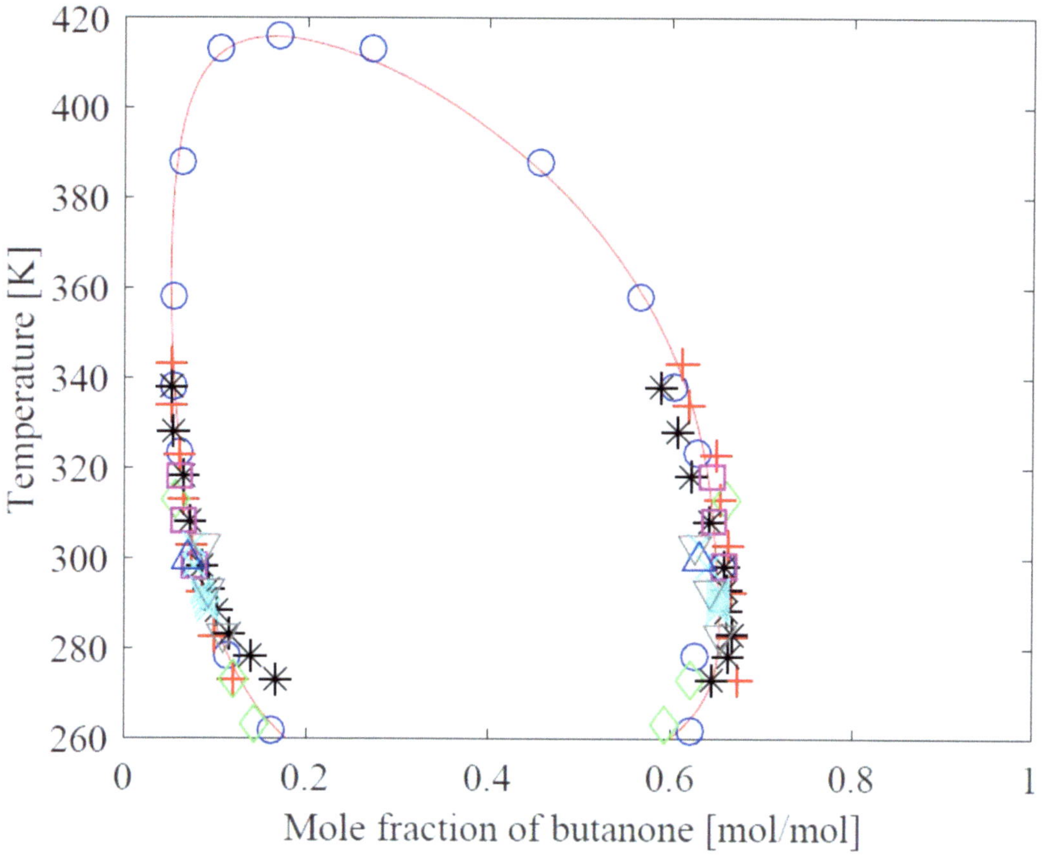

Fig. 4.5.: T-x diagram of the LLE butanone/water. Data are taken from Campbell et al. (18) (blue circle), Stephenson (81) (red plus), Ghildiyal et al. (30) (green diamond), Ferino et al. (27) (black asterisk), Arce et al. (4) (magenta square), Ruiz et al. (69) (cyan cross), Skrzec and Murphy (72) (blue upward triangle) and Dechambre et al. (22) (grey downward triangle). The mole fractions predicted with the parameters from Subsection 4.7.2 are plotted as a red line. The regression results in a good fit over the full temperature range in which the LLE of the butanone/water system exists.

4.7.4. Interpretation of Aspen Plus DRS results

For each case study presented in this chapter, the same measurement data are used for the regression in Aspen Plus as in BOARPET, and the Aspen Plus DRS regression tool completes without error messages in the status window. In the following, we will discuss the predictions of Aspen Plus DRS and three-phase flash. In Aspen Plus DRS, three-phase flash is automatically called as a consistency test. We have access to its results only through the history file and only in case an inconsistency is detected, i.e., the results of three-phase flash and of Aspen Plus DRS are not in agreement. In addition to three-phase flash, further solvers for flash problems are available in the unit operation block of Aspen Plus. However, only the latter need a feed as an input argument. In Chapter 5, the unit operation blocks DECANTER, FLASH3 and RGIBBS are tested.
In this subsection, four different errors that are associated with the following warning that is recorded in the report file of the regression of the LLE butanone/water are discussed. Note that the occurrence of this error might not be obvious to the standard user, as no error messages are shown in the status window.

- * ERROR: REGRESSION RESULTS DO NOT AGREE WITH 3-PHASE FLASH RESULTS.
  RERUN WITH DIFFERENT PARAMETER INITIAL GUESSES.

Accordingly, DRS and three-phase flash results are reported not to be in agreement for several data points in the history file of the LLE butanone/water. The predicted mole fractions from this file are presented in Table G.6.

The top left diagram in Figure 4.6 shows the surface of reduced Gibbs free energy of mixing at 263.34 K and the secant through the predicted compositions of Aspen Plus DRS at this temperature. Though the predicted compositions of DRS and three-phase flash differ significantly, it is difficult to discriminate the respective secants visually. Because of the low absolute curvature of a large part of the graph in the top left diagram in Figure 4.6 it seems likely that there is a temperature close to 263.34 K for which an unphysical phase split with infinitely many phases is predicted. Moreover, Aspen Plus DRS erroneously predicts phase compositions at 414.57 K which even violate a necessary condition of stability (isopotential), as can be seen in the top right diagram in Figure 4.6. Conclusively, the formulation used for parameter estimation by Aspen Plus DRS could be of the type described in Section 3.1. At the same temperature Aspen Plus three-phase flash erroneously predicts phase compositions that represent a local (as opposed to a global) minimum of the flash problem, i.e., the state is not stable, as shown in the bottom left diagram in Figure 4.6. The behaviour of the NRTL model at 415.71 K and other elevated temperatures corresponds to the category D as specified by Mitsos et al. (56). While three-phase flash predicts a stable homogeneous phase as a trivial solution of the LLE flash problem, Aspen Plus DRS predicts LLE compositions at points where the surface of Gibbs free energy of mixing is locally concave with respect to the mole fractions of both species in the binary system, as presented in the bottom right diagram in Figure 4.6. This corresponds to a local maximum of the corresponding flash problem, see e.g. Glass et al. (33). Both predictions vary significantly from the compositions measured by Campbell et al. (18) at 413.15 K and 415.75 K.

Altogether, as all observed predictions of the three-phase flash results in Aspen Plus regression locally fulfil strict convexity of the G^E-model at the predicted compositions, the formulation used in this algorithm is assumed to include 2nd-order sufficient KKT conditions as constraints. Still, three-phase flash does not represent a reliable stability test, as its predictions are observed to violate Baker's criterion. Aspen Plus DRS, on the other hand, is observed to violate a 1st-order necessary condition of stability (isopotential), as can be seen in the top right diagram in Figure 4.6. Hence, regression with Aspen Plus DRS can result in fits of diffusive equilibria with G^E or EoS models that violate thermodynamics. As a result, the obtained models cannot be used. In the following chapter, the consequences of yet using these models for prediction of diffusive equilibria are investigated by employing the unit operation blocks DECANTER, FLASH3 and RGIBBS in Aspen Plus, as well as different formulations of the flash problem implemented in GAMS (28).

From Figures G.1 to G.5 it can be observed that only the parameters regressed for the LLE butanone/water with Aspen Plus DRS allow for multiple double tangents to the Gibbs free energy function in the temperature range of the respective measurement data. Thus, this parameter set is used for comparison of different methods for solution of the flash problem in Chapter 5.

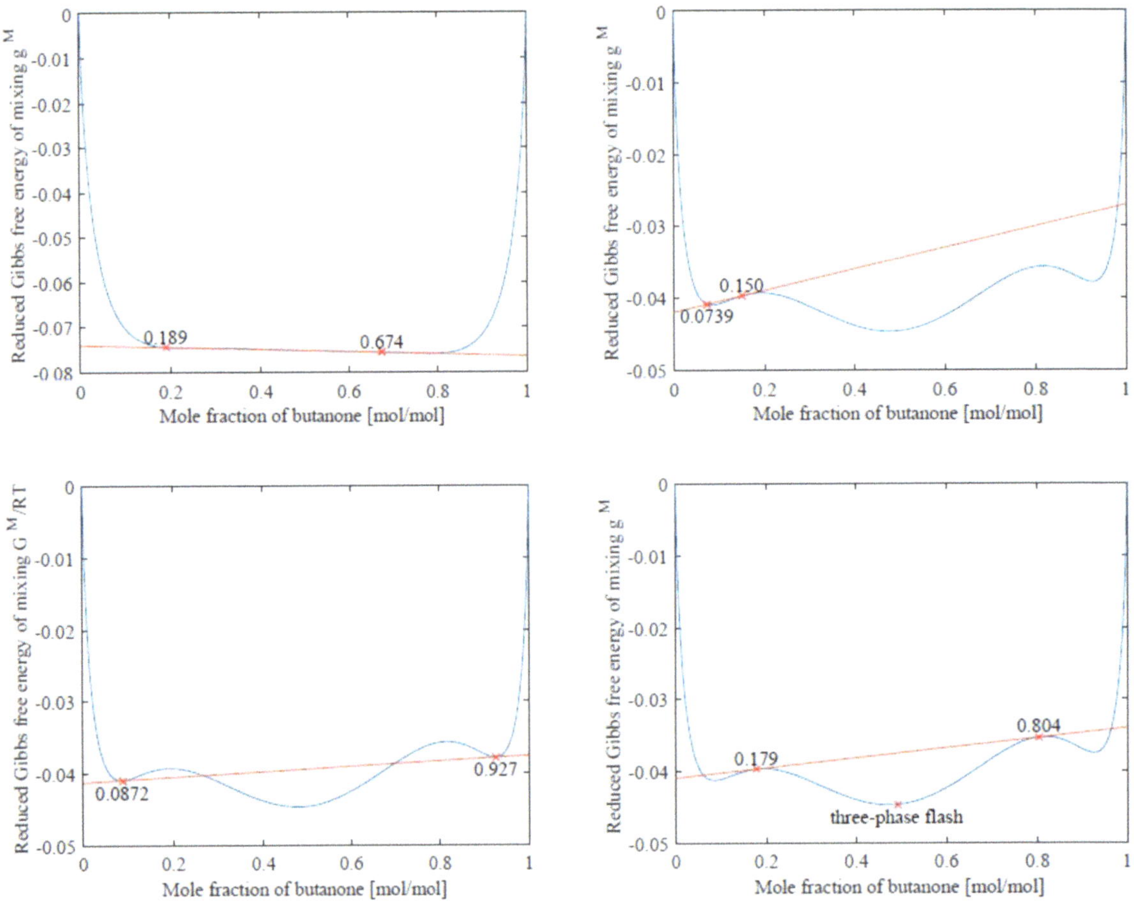

Fig. 4.6.: $g^M(x)$ diagrams for the LLE butanone/water. Top left: Prediction of Aspen Plus DRS at 263.34 K. It is not clear whether $g^M(x)$ is linear or not. If it is linear, this model predicts infinitely many phases, which is unphysical. Top right: The prediction of Aspen Plus DRS at 414.57 K violates the isoactivity condition. Bottom left: The prediction of three-phase flash at 414.57 K does not fulfil Baker's criterion. Hence, the results of three-phase flash should not be trusted to be a reliable test for the stability of Aspen Plus DRS predictions. Bottom right: While Aspen Plus DRS predicts a local maximum of the flash problem, three-phase flash predicts a trivial solution; however, both predictions vary significantly from the compositions measured by Campbell et al. (18) at 413.15 K and 415.75 K.

5. Flash simulations

In this chapter, different methods for flash simulations of binary LLE are tested, in order to assess the risks that may result from the violation of thermodynamic constraints during regression and/or during the prediction of diffusive equilibria, e.g., in the context of process simulation. Herein, the most challenging parameter sets of the case studies from Chapter 4 are employed. Four cases are discriminated in the evaluation of risks of predicting diffusive equilibria with regressed models:

1st The results of regression satisfy thermodynamic constraints and the global minimum is found for the resulting flash problem,

2nd the results of regression satisfy thermodynamic constraints and a solution that is significantly different from the globally optimal solution is found for the resulting flash problem,

3rd the results of regression violate thermodynamic constraints and the global minimum is found for the resulting flash problem,

4th the results of regression violate thermodynamic constraints and a solution that is significantly different from the globally optimal solution is found for the resulting flash problem.

Only in the 1st case of this risk evaluation, the predictions of the flash solver can be guaranteed to be consistent with the phase compositions that are predicted by the regression tool. If the results of regression represent a good fit and satisfy thermodynamic constraints, the global optimum of the flash problem (1st case) is the solution that is consistent with experimental evidence, and any solution that is significantly different from the globally optimal solution (2nd case) is not. If however, the results of regression violate thermodynamic constraints, any solver may predict compositions that are not consistent with experimental evidence. In this case, global methods will not give better guarantees than local methods, in the sense of flash results that agree with experimental evidence. More precisely, in the 3rd case and the

4th case, the prediction of the flash solver may for some feed be consistent with experimental evidence, but this cannot be guaranteed.

Two modelling environments are considered: The first is Aspen Plus V8.8 (7) and the second is the General Algebraic Modeling System (GAMS) Release 24.8.4 (28). In Aspen Plus the following unit operation blocks are considered:

- DECANTER,
- FLASH3,
- RGIBBS.

As some authors seem to suggest (48; 24; 53), an algorithm based on the work of Lucia and Xu (49) is likely implemented in the unit operation blocks RGIBBS and FLASH3 of Aspen Plus. However, in the help documentation of Aspen Plus V8.8 (7) Gautam and Seider (29), White and Seider (92) and Schott (71) are referenced regarding RGIBBS. For more general aspects of modelling of phase equilibria in Aspen Plus, the reader is referred to Boston (14) and Boston and Mathias (15). As a reference, the flash problem is also solved with the following global and local solvers through GAMS:

- BARON (global) (83),
- KNITRO (local) (5),
- SNOPT (local) (31; 32).

These solvers are used to test three different formulations of the flash problem implemented in GAMS. In Section 5.3, these formulations are presented. The NRTL model with parameter sets reported by the following researchers is used for testing of the different methods:

I this work (reported in Section G.5, discussed in Subsection 4.7.4),

II Glass et al. (33) (Table 3),

III Uusi-Kyyny et al. (88) (Table 8, Case 1),

IV Glass and Mitsos (37) (Table 3, Case GN),

V Glass and Mitsos (37) (Table 2, Case GN).

Note that these parameter sets are used because they allow for multiple local optima of Gibbs free energy at several feed compositions.

5.1. Methods and materials in Aspen Plus

The Inside-Out algorithm is used for all flash calculations in Aspen Plus except the calculations in RGIBBS, which uses the Gibbs algorithm as implemented in Aspen Plus. To achieve convergence, the maximal numbers of iterations are set to 9999 for both algorithms for the calculations with parameter set I. Apart from that, default settings are used for DECANTER, FLASH3 and RGIBBS. The options for flash calculations in Aspen Plus can not only be set individually for each block, but also globally in the tab "Flash Convergence" under "Calculation Options" in the "Setup" menu. Here also additional options which are not listed in the "Block Options" menu of an individual block, are available. In the present work, the defaults are used for these options.

5.2. Methods and materials in GAMS

Unless specified otherwise, the default option settings are used for GAMS and for the solvers BARON, KNITRO and SNOPT. For the initialisation of variables of the flash problems described in Section 5.3 the default setting of GAMS (28) is used, which uses zero, if this is within the constant bounds, or else the bound closest to zero as initial value. The only exception is the objective function value of Formulation 5.3, which is initialised with 1×10^{-10}. For the mole fraction variables, the lower bound x_{min} is set to $1 \times 10^{-5}\,\text{mol}\,\text{mol}^{-1}$ and the upper bound x_{max} is set to $1 - x_{min}$. The relative and absolute termination tolerances are set to 1×10^{-5} for all problems that are solved with GAMS in the present work. The objective function values and mole fractions in this chapter are reported with three significant digits, while the CPU time of the solvers is reported with two significant digits.

5.3. Formulations for the simulation of binary liquid-liquid equilibria

In this section, three formulations of the flash problem for binary LLE are presented. The first is based on the problem given by Equations (2.9a) - (2.9e). However, an implicit constraint on the molar phase fractions in that problem is now stated explicitly by Equation (5.1b), as box constraints may help some solvers to improve performance. Furthermore, tighter lower bounds

x_{min} and upper bounds x_{max} for the mole fraction variables $x_{i,j}$ are imposed by Equations (5.1c) and (5.1d) because the ideal Gibbs free energy of mixing given by Equation (2.16) is not defined for pure substances. The Formulation 5.1 then reads

$$\min_{\mathbf{x},\phi} \sum_{j\in J} \phi_j g^M(\mathbf{x}_j) \tag{5.1a}$$

subject to

$$\sum_{i\in I} x_{i,j} - 1 = 0, \text{ for all } j \in J, \tag{2.9b}$$

$$\sum_{j\in J} x_{i,j}\phi_j - x_i^0 = 0, \text{ for all } i \in I, \tag{2.9c}$$

$$\phi_j \geqslant 0, \text{ for all } j \in J, \tag{2.9d}$$

$$\phi_j \leqslant 1, \text{ for all } j \in J, \tag{5.1b}$$

$$x_{i,j} \geqslant x_{min}, \text{ for all } i \in I, j \in J, \tag{5.1c}$$

$$x_{i,j} \leqslant x_{max}, \text{ for all } i \in I, j \in J. \tag{5.1d}$$

As the scope of the present work is limited to binary LLE the index set of species I and the index set of phases J are both equal to $\{1, 2\}$ for all formulations in this chapter. In this case, the formulation contains 4 equality and 12 inequality constraints in 6 continuous variables. The second formulation used in this chapter expands the first formulation by imposing the isoactivity condition and convexity of the surface of Gibbs free energy at the predicted compositions. These conditions are necessary conditions for stability as pointed out in Chapters 2 and 3, respectively. By these means the feasible set is restricted to strict local minima and stationary points that correspond to the limit of stability, see e.g. Glass et al. (33). The Formulation 5.2

then reads

$$\min_{\mathbf{x},\phi} \sum_{j\in J} \phi_j g^M(\mathbf{x}_j) \tag{5.1a}$$

subject to

$$\sum_{i\in I} x_{i,j} - 1 = 0, \text{ for all } j \in J, \tag{2.9b}$$

$$\sum_{j\in J} x_{i,j}\phi_j - x_i^0 = 0, \text{ for all } i \in I, \tag{2.9c}$$

$$\phi_j \geqslant 0, \text{ for all } j \in J, \tag{2.9d}$$

$$\phi_j \leqslant 1, \text{ for all } j \in J, \tag{5.1b}$$

$$x_{i,j} \geqslant x_{min}, \text{ for all } i \in I, j \in J, \tag{5.1c}$$

$$x_{i,j} \leqslant x_{max}, \text{ for all } i \in I, j \in J, \tag{5.1d}$$

$$a_{i,1} = a_{i,2}, \text{ for all } i \in I, \tag{2.21}$$

$$\left.\frac{\partial^2 g^M}{\partial x^2}\right|_{x_{1,j}} \geqslant 0, \text{ for all } j \in J. \tag{5.2a}$$

Formulation 5.2 contains 6 equality constraints and 14 inequality constraints in 6 variables. The third formulation used in this chapter represents a completely different approach, motivated by the author's wish to reduce the number of independent variables to unity. Reducing the number of variables, however, seems to come at the cost of introducing an iterative procedure. More precisely, an iterative search for the mole fraction with maximal violation of the supporting tangent criterion from Baker et al. (8) is conducted and the secant is updated with the optimised mole fraction after every iteration. Hence, the objective function is the difference in reduced Gibbs free energy between the hypersurface and the hyperplane , i.e., the negative of the tangent plane distance function as described by Michelsen (54). The problems solved in the iterations contain one variable with constant lower and upper bounds and no equality constraints. They are of the following form:

$$\max_{x_{1,iter}} \left[g^M(x_{1,2}) + \frac{g^M(x_{1,1}) - g^M(x_{1,2})}{x_{1,1} - x_{1,2}} \cdot (x_{1,iter} - x_{1,2}) - g^M(x_{1,iter}) \right] \tag{5.3a}$$

$$\text{subject to } x_{min} \leqslant x_{1,iter} \leqslant x_{max}. \tag{5.3b}$$

Herein, $x_{1,iter}$ denotes the mole fraction at which the Gibbs free energy of mixing and the secant are evaluated. The solution $x^*_{1,iter}$ is the mole fraction of species 1 of maximal violation of Baker's criterion. The mole fractions of the two predicted phases are initialised as $x_{1,1} = x_{max}$ and $x_{1,2} = x_{min}$. Note that the mole fractions of the predicted phases must be initialised with a feasible point of the flash problem with Formulation 5.1. After every iteration the solution $x^*_{1,iter}$ is used to update the composition of one predicted phase as follows:

$$\begin{aligned} &\text{if } x^*_{1,iter} \geqslant x^0_1 \\ &\quad \text{then } x_{1,1} := x^*_{1,iter} \\ &\quad \text{and } x_{2,1} := 1 - x^*_{1,iter} \\ &\quad \text{and update reduced Gibbs free energy of mixing of phase 1} \\ &\text{else } x_{1,2} := x^*_{1,iter} \\ &\quad \text{and } x_{2,2} := 1 - x^*_{1,iter} \\ &\quad \text{and update reduced Gibbs free energy of mixing of phase 2.} \end{aligned} \tag{5.4}$$

The problem resulting from Formulation 5.3 and the updated predictions for phase 1 and 2 is then solved repeatedly followed by the scheme described in Equation (5.4) as long as the maximal value found for Equation (5.3a) is greater equal a tolerance of 1×10^{-10}. It is important to point out, that Formulation 5.3 does not guarantee the isopotential condition to be satisfied and convergence to the global optimum is not proven for the combination of Formulation 5.3 with the scheme described in Equation (5.4).

In order to provoke stationary points of Formulation 5.1 other than minima, two variations of Formulation 5.2 are used, see e.g. Glass et al. (33). Since methods and solvers in Aspen seem to erroneously report these, see Section 4.7.4, it is of interest to find such points with the reference problems in GAMS, in order to assess potential risks during, e.g., process simulation. Thus, the aim of employing the following formulations is to find the worst-case scenario when 2nd-order KKT conditions are not used by the flash solver. In these variations, all equations contained in Formulation 5.2 except Equation (5.2a) are left unchanged. A local maximum or a solution corresponding to the limit of thermodynamic stability is found, if the constraint expressed by Equation (5.2a) is replaced by

$$\left.\frac{\partial^2 g^M}{\partial x^2}\right|_{x_{1,j}} \leqslant 0, \text{ for all } j \in J. \tag{5.5}$$

This formulation is called Formulation 5.2b. If in addition to fulfilling the isoactivity condition the surface of Gibbs free energy is strictly convex at one predicted composition and strictly concave at the other, the stationary point is a saddle as opposed to a global minimum. However only less equal and greater equal can be implemented as inequalities in GAMS. Thus, the constraints are relaxed by imposing the surface of Gibbs free energy to be convex at the one predicted composition and concave at the other. This formulation is called Formulation 5.2c. Further flash formulations that are not employed in the present work can be found in Appendix H.

5.4. Parameter set I

As discussed in Subsection 4.7.4, the NRTL model obtained for the LLE butanone/water with the parameters from Section G.5 violates thermodynamic constraints. Hence, the 3rd and 4th cases of the risk evaluation will be discussed in this section.

The $g^M(\mathbf{x}_j)$ functions obtained with the parameters from Section G.5 are especially challenging at temperatures near the lower and upper limits of the regressed temperature range. At the lowest predicted temperature the curvature of the $g^M(\mathbf{x}_j)$ function is near zero over more than half of the mole fraction range, as can be seen in top left diagram in Figure 4.6. This topology may be challenging for solvers because infinitely many sets of predicted compositions fulfil the optimality criteria if too high tolerances are selected. The unphysical prediction of infinitely many phases as solution of a flash problem can be the optimal solution, only if $g^M(x)$ has undulation points. However, the adjustment of tolerances and the prediction of diffusive equilibria with more than two phases are beyond the scope of the present work and thus, a temperature near the upper limit of the regressed temperature range is selected for evaluation of risks.

At $414.57\,\mathrm{K}$, the prediction of Aspen Plus DRS violates 1st-order necessary conditions for optimality like the isoactivity condition (see top right diagram in Figure 4.6) thus, in the 3rd case, the prediction of the flash solver will differ significantly from the prediction of Aspen Plus DRS for feeds that lie between the phase compositions of this prediction.

Feed compositions with the following set of mole fractions of butanone in water are tested: $F_1 = \{0.05\,\mathrm{mol\,mol^{-1}}, 0.2\,\mathrm{mol\,mol^{-1}}, 0.5\,\mathrm{mol\,mol^{-1}}, 0.8\,\mathrm{mol\,mol^{-1}}\}$. The conditions for the flash calculations with unit operation blocks in Aspen Plus are set to $414.57\,\mathrm{K}$ and $25\,\mathrm{bar}$ so that the vapour fraction is zero. FLASH3, DECANTER and RGIBBS report a trivial solution for the first feed composition in set F_1, which corresponds to the 3rd case. Moreover, DECANTER reports a local minimum, which corresponds to the 4th case, with the same predicted compositions for all remaining feed compositions in set F_1, as presented in Figure 5.1. In contrast to that, FLASH3 and RGIBBS find the global minimum for the second feed

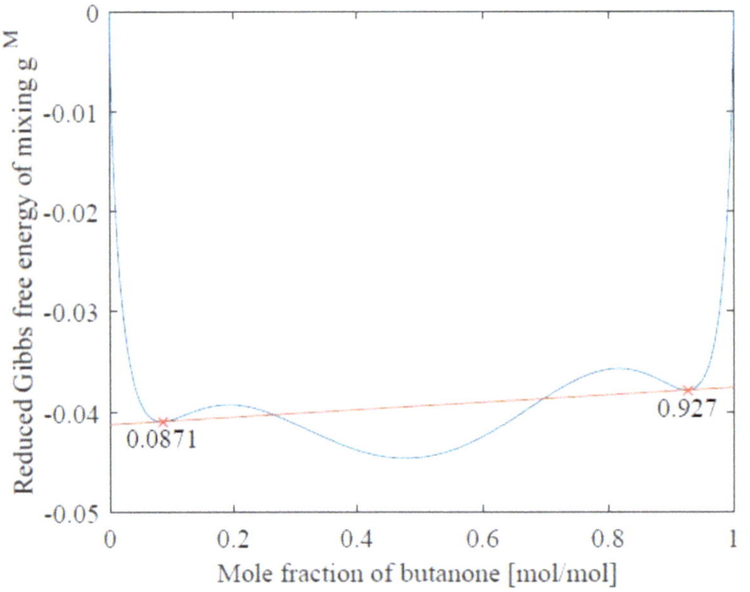

Fig. 5.1.: $g^M(x)$ for the parameters from Section G.5 and the secant at the compositions predicted by DECANTER at 414.57 K and 25 bar. DECANTER wrongly predicts a local minimum (4th case) of the flash problem for all feed compositions in set F_1 except the first. Aspen Plus DRS predicts $x_{1,1} = 0.150$ mol mol^{-1} and $x_{1,2} = 0.0739$ mol mol^{-1}.

compositions in set F_1 and the trivial solution for the third feed compositions in set F_1, which correspond to the 3rd case. The former is depicted in Figure 5.2. While FLASH3 predicts the same local minimum for the fourth feed compositions in set F_1 as DECANTER (4th case), RGIBBS predicts the global minimum that is depicted in Figure 5.3 for this feed (3rd case). The three flash problems described in Sections 5.3 are solved in GAMS for the same set of feed compositions F_1. The secants through the predicted compositions are depicted in Figure 5.4 for some of the flash problems that are solved in GAMS based on parameter set I. For the fourth feed composition in set F_1, BARON predicts the global minimum of the given flash problem with Formulation 5.1 and the objective function value -0.0399 in 15 s CPU time. SNOPT predicts the same compositions and objective function value in 0.13 s CPU time. For the second feed composition in set F_1, SNOPT predicts a local minimum of the given flash problem with Formulation 5.1 and the objective function value -0.0405 in 0.094 s CPU

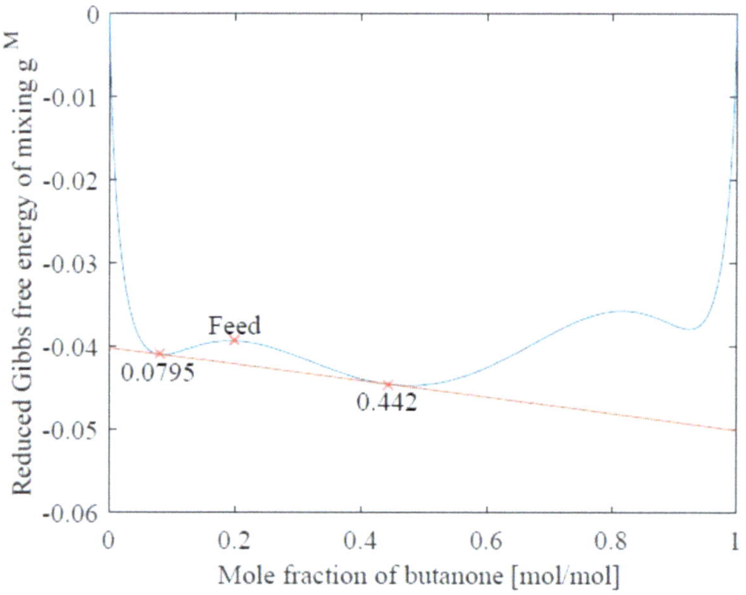

Fig. 5.2.: $g^M(x)$ for the parameters from Section G.5 and the secant at the compositions predicted by FLASH3 and RGIBBS at 414.57 K, 25 bar for the second feed composition in set F_1. This solution is the global minimum (3rd case) of the flash problem. Aspen Plus DRS predicts $x_{1,1} = 0.150$ mol mol^{-1} and $x_{1,2} = 0.0739$ mol mol^{-1}.

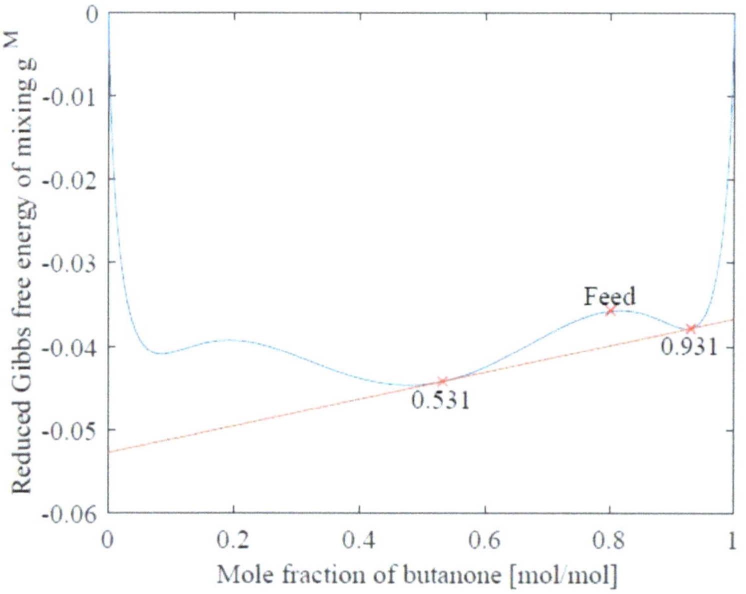

Fig. 5.3.: $g^M(x)$ for the parameters from Section G.5 and the secant at the compositions predicted by RGIBBS at 414.57 K, 25 bar for the fourth feed composition in set F_1. This solution is the global minimum (3rd case) of the flash problem. Aspen Plus DRS predicts $x_{1,1} = 0.150$ mol mol^{-1} and $x_{1,2} = 0.0739$ mol mol^{-1}.

time. For the first feed composition in set F_1, a solution violating the isoactivity condition is returned by SNOPT with 5 iterations after 0.046 s CPU time. The same solution is returned by KNITRO with the same number of iterations after 0.47 s CPU time. The value of Equation (5.1a) for this solution is -0.0443. The first four solutions for the iterative maximisation of the violation of Baker's criterion are marked with the numbers 1 to 4 in the bottom right diagram in Figure 5.4, while the secants visualise how Equation (5.4) updates the predicted phase compositions using the solution of each iteration. Finally, BARON returns the trivial solution of this problem after 15 iterations and 1.0 s CPU time.

In case Formulation 5.2b is used, a local maximum of the flash problem as presented in the top diagram of Figure 5.5 is found by BARON with objective function value -0.0375 in 5.3 s CPU time. This maximum (4th case) fulfils the isoactivity condition (Equation (2.21)) and could also be return by other solvers based on a formulation that does not impose convexity of the surface of Gibbs free energy at the predicted phase compositions (Equation (5.2a)). The bottom diagrams of Figure 5.5 present two saddles that are obtained by BARON, when Formulation 5.2c is used. The saddle presented in the bottom left is obtained with objective function value -0.0401 in 3.9 s CPU time, while the saddle presented in the bottom right diagram is obtained with objective function value -0.0381 in 5.6 s CPU time. These saddles (4th case) fulfil the isoactivity condition (Equation (2.21)) and could also be return by other solvers based on formulations that do not impose convexity of the surface of Gibbs free energy at the predicted phase compositions (Equation (5.2a)).

In a nutshell, only the unit operation block RGIBBS works reliably (3rd case) in all cases tested for parameter set I. Both, DECANTER and FLASH3 report a local (as opposed to a global) minimum (4th case) for at least one of the feed compositions tested for parameter set I.

5.5. Parameter set II

Glass et al. (33) report that regression of binary interaction parameters of the NRTL model in Aspen Plus V8.8 (7) followed by solving the flash problem with RGIBBS can lead to erroneous predictions of phase equilibria. The following parameters are obtained by Glass et al. (33) with Aspen Plus for the LLE of 2-MTHF (1) and water (2), and used in this section to evaluate the risk of obtaining different predictions for the flash problem than those used by the regression

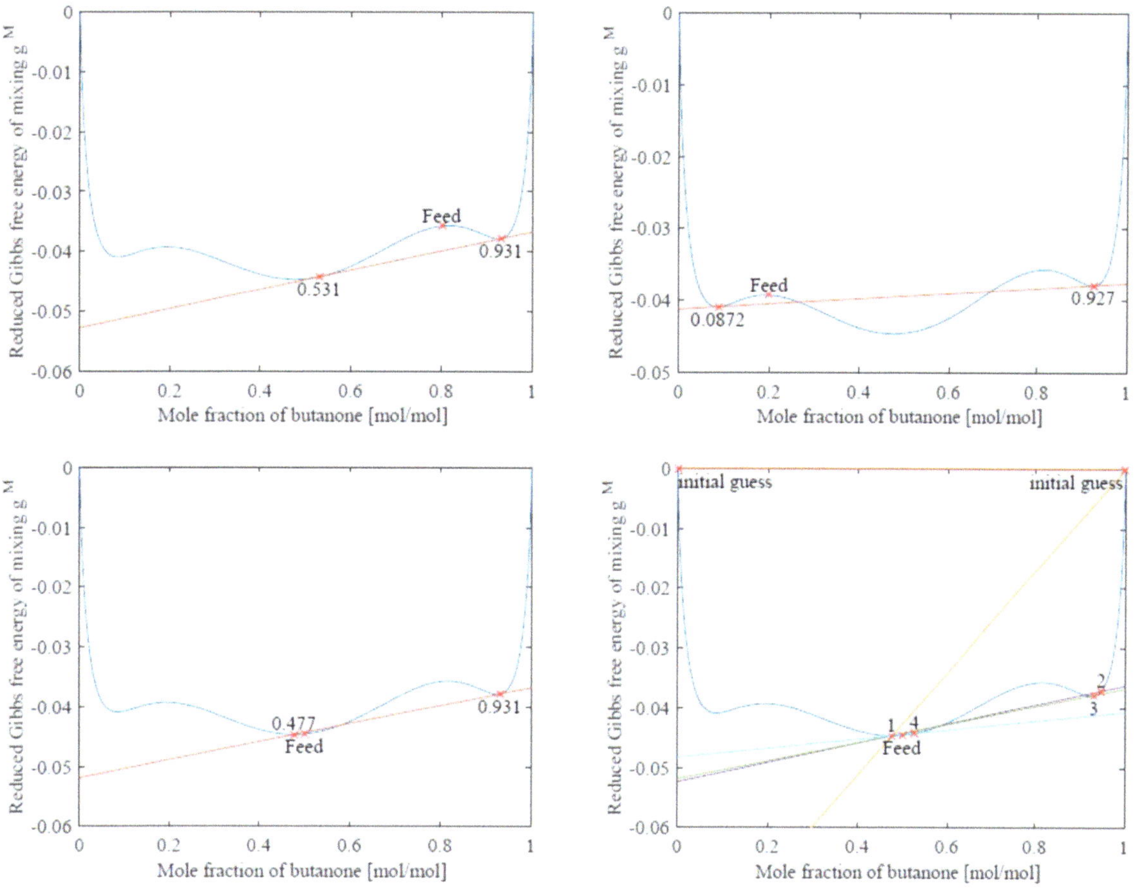

Fig. 5.4.: $g^M(x)$ diagrams for the parameters from Section G.5 at 414.57 K. Top left: Secant at the compositions predicted by SNOPT and BARON at the fourth feed composition in set F_1 for Formulation 5.1 (3rd case). Top right: Secant at the compositions predicted by SNOPT at the second feed composition in set F_1 for Formulation 5.1 (4th case). Bottom left: Secant at the compositions predicted by KNITRO and SNOPT at the third feed composition in set F_1 for Formulation 5.3. This solution violates the isoactivity condition (Equation (2.21)), which is a 1st-order necessary condition for optimality (4th case). Bottom right: Secants of iterations 1 to 4 predicted by BARON at the third feed composition in set F_1 for Formulation 5.3. Finally, the solver converges to the trivial solution after 15 iterations and 1.0 s CPU time (3rd case). The mole fractions $x_{1,1} = 0.150\,\mathrm{mol\,mol^{-1}}$ and $x_{1,2} = 0.0739\,\mathrm{mol\,mol^{-1}}$ are predicted by Aspen Plus DRS.

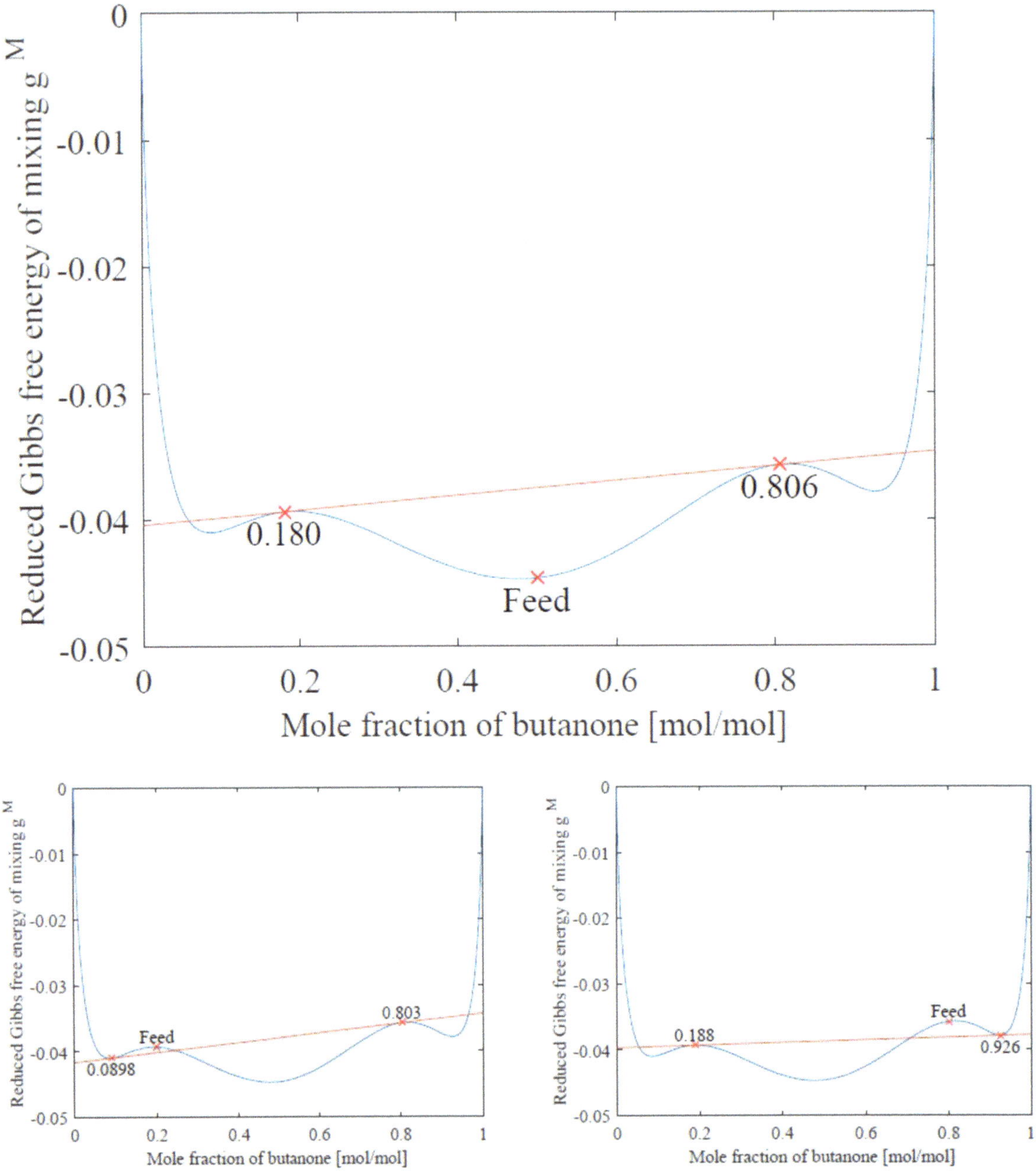

Fig. 5.5.: $g^M(x)$ diagrams for the parameters from Section G.5 at 414.57 K. Top: Local maximum (4th case) predicted by BARON at the third feed composition in set F_1 for Formulation 5.2b. Bottom: Saddles (4th case) predicted by BARON for Formulation 5.2c. The depicted maximum and the saddles fulfil the isoactivity condition (Equation (2.21)) and could also be return by other solvers based on formulations that do not impose convexity of the surface of Gibbs free energy at the predicted phase compositions (Equation (5.2a)). The mole fractions $x_{1,1} = 0.150\,\text{mol}\,\text{mol}^{-1}$ and $x_{1,2} = 0.0739\,\text{mol}\,\text{mol}^{-1}$ are predicted by Aspen Plus DRS.

tool:

$$
\begin{aligned}
a_{12}^{Aspen} &= -31.4449 && \textit{regressed}\\
b_{12}^{Aspen} &= 1079.95 && \textit{regressed}\\
f_{12}^{Aspen} &= 0 && \textit{fixed}\\
e_{12}^{Aspen} &= 5.26904 && \textit{regressed}\\
c_{12}^{Aspen} &= 0.405204 && \textit{regressed}\\
d_{12}^{Aspen} &= 0 && \textit{fixed}\\
a_{21}^{Aspen} &= 142.132 && \textit{regressed}\\
b_{21}^{Aspen} &= -7712.72 && \textit{regressed}\\
f_{21}^{Aspen} &= 0 && \textit{fixed}\\
e_{21}^{Aspen} &= -19.8342 && \textit{regressed}
\end{aligned}
$$

These parameters are regressed violating thermodynamic constraints. Hence, the 3rd and 4th cases of the risk evaluation will be discussed in this section. Glass et al. discuss the deviation between the regressed compositions and the compositions predicted by RGIBBS at $423.176\,\mathrm{K}$. The same temperature is selected for the evaluation of risks in the present work. Feed compositions with the following set of mole fractions of 2-MTHF in water are tested in Aspen Plus and GAMS: $F_2 = \{0.2\,\mathrm{mol\,mol^{-1}}, 0.8\,\mathrm{mol\,mol^{-1}}, 0.95\,\mathrm{mol\,mol^{-1}}\}$.

In contrast to the results for parameter set I, DECANTER returns the global minimum for all tested feed compositions for parameter set II (3rd case), but RGIBBS does not, see Figure 5.6. The phase compositions of the global minima for the first two feed compositions in set F_2 are presented in the top left diagram in Figure 5.6. For the second feed composition in set F_2, DECANTER and FLASH3 report the global minimum, while for the first feed composition in set F_2 only DECANTER reports the global minimum (4th case). On the other hand, RGIBBS predicts local minima (4th case) for the first two feed compositions in set F_2, as depicted in the top right diagram and the middle left diagram in Figure 5.6. For the first feed composition in set F_2, FLASH3 predicts the same local minimum (4th case) as RGIBBS. For the last feed composition in F_2, DECANTER, FLASH3 and RGIBBS report the trivial solution, which is the global minimum in this case (3rd case). The three flash problems described in Sections 5.3 are solved in GAMS for the same set of feed compositions F_2. For Formulation 5.1 and the first feed composition in set F_2, both BARON and SNOPT report the global minimum (3rd case) presented in the middle right diagram in Figure 5.6 with the objective function value -0.0143. The CPU time is $2.7\,\mathrm{s}$ for BARON and $0.094\,\mathrm{s}$ for SNOPT. For Formulation 5.1 and the second feed composition in set F_2 SNOPT reports the local minimum (4th case) presented in the middle left diagram in Figure 5.6 with the objective function value -0.0288 after $0.063\,\mathrm{s}$ CPU time, while BARON reports the global minimum (3rd case). In case Formulation 5.2b is used and the absolute termination tolerance is set to 1×10^{-4}, a local maximum (4th case) of the flash problem as presented in the bottom left diagram in Figure 5.6 is found by BARON for the second feed composition in set F_2. The bottom right diagram in Figure 5.6 and the left diagram in Figure 5.7 present two saddles (4th case) that are obtained by BARON, when Formulation 5.2c is used. The maximum and the saddles fulfil the isoactivity condition (Equation (2.21)) and could also be return by other solvers based on formulations that do not impose convexity of the surface of Gibbs free energy at the predicted phase compositions (Equation (5.2a)). Of all predictions presented in Figures 5.6 and 5.7, the predicted compositions of the saddle (4th case) depicted in the bottom right diagram in Figure 5.6 is closest to the compositions measured by Glass et al. at $423.15\,\mathrm{K}$, as this saddle corresponds to the prediction erroneously used by Aspen Plus DRS. Thus, a prediction that is consistent with experimental evidence can be obtained, if both the regression tool and the flash solver return the same solution violating thermodynamic constraints. However, flash solvers that guarantee satisfaction of thermodynamic constraints will return a significantly different prediction. Moreover, even the same solver base on the same regressed model may return solutions that are not consistent with experimental evidence for a different feed. The secant through the phase compositions that are predicted for the second feed composition in set F_2 with Formulation 5.3 employing BARON is presented in the right diagram in Figure 5.7. Interestingly, based on Formulation 5.3, BARON reports a solution that is violating the isoactivity condition at this feed composition (4th case). In Appendix I, the results reported for this flash problem by different global solvers in GAMS (28) are discussed.

In a nutshell, only the unit operation block DECANTER works reliably (3rd case) in all cases tested for parameter set II. Both, RGIBBS and FLASH3 report a local (as opposed to a global) minimum (4th case) for at least one of the feed compositions tested for parameter set II.

5.6. Parameter set III

Uusi-Kyyny et al. (88) report two NRTL parameter sets with four and six regressed parameters for the LLE of 2-MF (1) and water (2). The values $298\,\mathrm{K}$ and $415\,\mathrm{K}$ are reported [ibid.] as temperature boundaries in which these models are valid. Both parameter sets imply two possible positions for double tangents to $g^M(x)$ at temperatures near the lower bound of the specified temperature range. The parameter set with 4 regressed parameters is selected for the flash simulations in this section:

$$
\begin{aligned}
a_{12}^{Aspen} &= -14.617 && \textit{regressed} \\
b_{12}^{Aspen} &= 6291.575 && \textit{regressed} \\
f_{12}^{Aspen} &= 0 && \textit{fixed} \\
e_{12}^{Aspen} &= 0 && \textit{fixed} \\
c_{12}^{Aspen} &= 0.3 && \textit{fixed} \\
d_{12}^{Aspen} &= 0 && \textit{fixed} \\
a_{21}^{Aspen} &= 2.912 && \textit{regressed} \\
b_{21}^{Aspen} &= 1031.634 && \textit{regressed} \\
f_{21}^{Aspen} &= 0 && \textit{fixed} \\
e_{21}^{Aspen} &= 0 && \textit{fixed}
\end{aligned}
$$

At $300\,\mathrm{K}$ $g^M(x)$ has two positions for possible double tangents between the lower and upper limits of mole fractions x_{min} and x_{max} specified in Section 5.1. As the resulting NRTL model satisfies thermodynamic constraints at this temperature, the 1st and 2nd cases of the risk evaluation are discussed in this section. Feeds with the following mole fractions of 2-MF in water are tested: $F_3 = \{0.1\,\mathrm{mol\,mol^{-1}}, 0.5\,\mathrm{mol\,mol^{-1}}\}$.

DECANTER and FLASH3 report the global minima (1st case) with the phase compositions depicted in the left diagram in Figure 5.8 for both feed compositions in set F_3, while RGIBBS reports the global minimum only for the first feed composition in set F_3. For the second feed composition in this set, RGIBBS reports the trivial solution (2nd case). The three flash problems described in Sections 5.3 are solved in GAMS with parameter set III for F_3. Some of the predictions are presented in the left diagram in Figure 5.8. Using Formulation 5.1, the global minimum for the first feed composition in set F_3 is found by BARON and SNOPT within $0.48\,\mathrm{s}$ and $0.078\,\mathrm{s}$ CPU time, respectively, while the global minimum for the second feed composition in set F_3 is found by BARON and SNOPT within $0.42\,\mathrm{s}$ and $0.062\,\mathrm{s}$ CPU time, respectively. Using Formulation 5.2, the global minima for the first and second feed compositions are found by BARON within $0.81\,\mathrm{s}$ and $0.56\,\mathrm{s}$ CPU time, respectively. Using Formulation 5.3, the global minima of the flash problems for the first and second feed compositions are found by BARON in 4 iterations after $0.17\,\mathrm{s}$ and $0.14\,\mathrm{s}$ CPU time, respectively.

In case Formulation 5.2b is used and the absolute termination tolerance is set to 1×10^{-4}, the global maximum (2nd case) of the flash problem as presented in the right diagram in Figure 5.8 is found by BARON for the second feed composition in set F_3 . This solution fulfils the isoactivity condition (Equation (2.21)) and could also be returned by other solvers for a formulation that does not impose convexity of the surface of Gibbs free energy at the predicted phase compositions (Equation (5.2a)).

In a nutshell, only the unit operation blocks DECANTER and FLASH3 work reliably (1st case) in all cases tested for parameter set III. RGIBBS reports a trivial solution (as opposed to a global minimum) (4th case) for one of the feed compositions tested for parameter set III.

5.7. Parameter sets IV and V

Glass and Mitsos (37) report that regression of binary interaction parameters of the NRTL model with the Gauss-Newton algorithm in DECHEMA DPP (23) can lead to prediction of spurious phase splits. The following parameters are obtained by Glass and Mitsos (37) with the Gauss-Newton algorithm in DECHEMA DPP and used in this section to give an example of a further commercially available regression tool that predicts erroneous numbers of phase splits. The parameters that are obtained by Glass and Mitsos (37) for the LLE of the binary system n-butanol (1) and water (2) read

Set IV

$$
\begin{aligned}
a_{12}^{Aspen} &= 1.6189395 && \textit{regressed}\\
b_{12}^{Aspen} &= 289.59486 && \textit{regressed}\\
f_{12}^{Aspen} &= 0 && \textit{fixed}\\
e_{12}^{Aspen} &= 0 && \textit{fixed}\\
c_{12}^{Aspen} &= 0.4362121 && \textit{regressed}\\
d_{12}^{Aspen} &= 0 && \textit{fixed}\\
a_{21}^{Aspen} &= 1.9835384 && \textit{regressed}\\
b_{21}^{Aspen} &= 139.84128 && \textit{regressed}\\
f_{21}^{Aspen} &= 0 && \textit{fixed}\\
e_{21}^{Aspen} &= 0 && \textit{fixed}.
\end{aligned}
$$

while the parameters that are obtained by Glass and Mitsos (37) for the LLE of the binary system tetrahydrofuran (THF) (1) and water (2) read

Set V

$$
\begin{aligned}
a_{12}^{Aspen} &= -1.4903631 && \textit{regressed}\\
b_{12}^{Aspen} &= 5210.0138000 && \textit{regressed}\\
f_{12}^{Aspen} &= -0.0234748 && \textit{regressed}\\
e_{12}^{Aspen} &= 0 && \textit{fixed}\\
c_{12}^{Aspen} &= 0.4273386 && \textit{regressed}\\
d_{12}^{Aspen} &= 0 && \textit{fixed}\\
a_{21}^{Aspen} &= 7.4404315 && \textit{regressed}\\
b_{21}^{Aspen} &= 311.8801700 && \textit{regressed}\\
f_{21}^{Aspen} &= -0.0149611 && \textit{regressed}\\
e_{21}^{Aspen} &= 0 && \textit{fixed}.
\end{aligned}
$$

For both parameter sets, feed compositions with the following set of mole fractions of species 1 are used to predict the corresponding phase splits: $F_4 = \{0.25\,\text{mol mol}^{-1}, 0.75\,\text{mol mol}^{-1}\}$. The regressed models of both binary systems are evaluated at $373.15\,\text{K}$. The following solutions are obtained with BARON based on formulation 5.2 (3rd case). For the flash simulation with parameter set IV and the first feed composition in set F_4 a solution with the absolute termination tolerance 1×10^{-4} and the objective function value -0.0687 and the mole fraction variables $x_{1,1} = 0.311\,\text{mol mol}^{-1}$ and $x_{1,2} = 0.225\,\text{mol mol}^{-1}$ is obtained in $15\,\text{s}$ CPU time, while for the second feed composition in set F_4 a solution with the absolute termination tolerance 1×10^{-5} and the objective function value -0.0667 and the mole fraction variables $x_{1,1} = 0.824\,\text{mol mol}^{-1}$ and $x_{1,2} = 0.631\,\text{mol mol}^{-1}$ is obtained in $8.2\,\text{s}$ CPU time. Similarly, for the flash simulation with parameter set V and the first feed composition in set F_4 a solution with the objective function value -0.0567 and the mole fraction variables $x_{1,1} = 0.261\,\text{mol mol}^{-1}$ and $x_{1,2} = 0.136\,\text{mol mol}^{-1}$ is obtained in $11\,\text{s}$ CPU time, while for the second feed composition in set F_4 a solution with the objective function value -0.0331 and the mole fraction variables $x_{1,1} = 0.984\,\text{mol mol}^{-1}$ and $x_{1,2} = 0.506\,\text{mol mol}^{-1}$ is obtained in $3.4\,\text{s}$ CPU time. Hence, both models discussed in the present section predict two phase splits at the tested temperature, though only one phase split is reported for the fluid systems n-butanol/water and THF/water by Sørensen and Arlt (76). The phase compositions reported [ibid.] for LLE at $373.15\,\text{K}$ are $x_{1,1} = 0.322\,\text{mol mol}^{-1}$ and $x_{1,2} = 0.0237\,\text{mol mol}^{-1}$ for n-butanol/water and $x_{1,1} = 0.389\,\text{mol mol}^{-1}$ and $x_{1,2} = 0.0888\,\text{mol mol}^{-1}$ for THF/water.

In a nutshell, with the Gauss-Newton algorithm in the DECHEMA Data Preparation Package regression results that predict erroneous numbers of phase splits can be obtained. As a consequence, results of flash simulations based on regressed models can be inconsistent with experimental evidence.

5.8. Discussion and comparison of methods

Summing up the most relevant observations regarding the binary LLE flash problem, the 2nd and 4th cases of those defined at the beginning of this chapter are avoided, if a global solver is employed for the solution of the flash problem. Thus, it is highly recommended by the author to use global solvers, and based on the observations of the present work this excludes the unit operation blocks of Aspen Plus. The use of global solvers is especially important, if $g^M(x)$ has more than one possible position for a double tangent, as erroneous prediction of trivial solutions may be avoided by a reliable initialisation method for a local flash solver. Generalised for arbitrary numbers of phases and species this means that employing global solvers is important, if there are at least two positions for a tangent hyperplane. In these cases, the author recommends using global solvers, e.g. BARON, in modelling environments like GAMS (28) or AMPL (2).

As none of the three unit operation blocks considered in Aspen Plus (DECANTER, FLASH3, RGIBBS) returned the global minimum for every flash problem tested, we conclude that local solvers are implemented in the "Inside-Out" algorithm and the "Gibbs" algorithm, which are employed for flash convergence in the unit operation blocks in Aspen Plus. The results of the unit operations models tested in Aspen Plus are qualitatively most similar to the results of the local solver SNOPT for Formulation 5.1. In addition, as the predictions returned by the tested unit operation models in Aspen Plus always fulfilled the isoactivity condition and convexity of $g(x)$ at the predicted compositions, formulations including Equations (2.21) and (5.2a) might be used in these models. However, the predictions of DECANTER, FLASH3 and RGIBBS have to be treated with caution, if a nonconvex flash problem is solved. In other unit operations that contain a sequence of flash problems, the error that is introduced by predictions of the flash solver which are not consistent with experimental evidence, will propagate.

Global and local solvers are used through GAMS with different formulations of the flash problem as a reference for the unit operation models in Aspen Plus. However, only the employed global solver BARON can be recommended as an alternative, as both local solvers considered in GAMS reported local infeasibility for several problems based on Formulation 5.2. The solution of the flash problem based on Formulation 5.2 with these solvers would benefit from an initialisation with a feasible point. If BARON is employed, it is observed that some of the problems based on Formulations 5.1 and 5.2 do not converge to the termination tolerances specified in Section 5.2 within $12\,000\,\mathrm{s}$ ($\approx 3\,\mathrm{h}$) CPU time. However, if the absolute termination tolerance would be 1×10^{-4}, all flash problems executed with BARON based on Formulations 5.1 and 5.2 would converge to the global minimum in under $80\,\mathrm{s}$ CPU time. Hence, the 2nd case and the 4th case of the risk evaluation as defined at the beginning of this chapter can be excluded when employing BARON based on Formulation 5.1 or Formulation 5.2, and the use of these formulations in combination with a global solver and appropriate termination tolerances is recommended. Since Formulation 5.3 cannot give better guarantees than the Formulations 5.1 and 5.2, it is not further discussed.

Different inequality constraints for second derivatives of g^M are used in Formulations 5.2b and 5.2c because we want to illustrate what trouble violation of convexity of the surface of Gibbs free energy at the predicted compositions can cause. The overall goal is to assess the risks to obtain erroneous predictions for flash problems. As several maxima and saddles are observed using Formulations 5.2b and 5.2c, it is clearly insufficient to find a solution of the flash problem satisfying 1st-order necessary KKT conditions, if several stationary points exist. Note that the phase compositions that correspond to the different stationary points differ significantly in the examples discussed in Sections 5.4 to 5.6. To avoid obtaining a maximum or a saddle as a solution for the flash problem, using solvers that fulfil 2nd-order sufficient KKT conditions is recommended. The unit operation models tested in Aspen Plus are observed to fulfil these conditions. However, it is insufficient for obtaining the global minimum to find a solution of the flash problem satisfying 2nd-order sufficient KKT conditions, if several minima exist.

Finally, the flash simulation must be evaluated in combination with the employed model for the fluid system. If the regression is violating thermodynamic stability, a local solver for the flash problem may still return the same predictions as the regression tool does (4th case). These predictions may be consistent to experimental evidence, but they will not be reproducible with a global solver (3rd case) for all feasible feed compositions and they may not be reproducible by other local solvers as well. Hence, employing a regression tool that guarantees thermodynamic stability and correct numbers of phase splits and phases per phase split is crucial for prediction of diffusive equilibria. As discussed in Section 4.7.4, this excludes the Aspen Plus Data Regression System. If, on the other hand, the regression is conducted guaranteeing those criteria and converging to global optimality, the global minimum of the flash problem (1st case) is for any feasible feed composition the solution that is consistent with the regressed measurement data and any solution with significant deviation from the globally optimal solution (2nd case) is not. As a result, it is strongly recommended to resort to global solvers for the solution of the flash problem. Based on the observations discussed in this chapter, this excludes the unit operation models in Aspen Plus. In case global solvers are employed for the solution of the flash problem, it is recommended to impose 2nd-order sufficient KKT conditions only if these constraints improve the convergence, as they are not sufficient conditions for thermodynamic stability, but global optimality of this problem is.

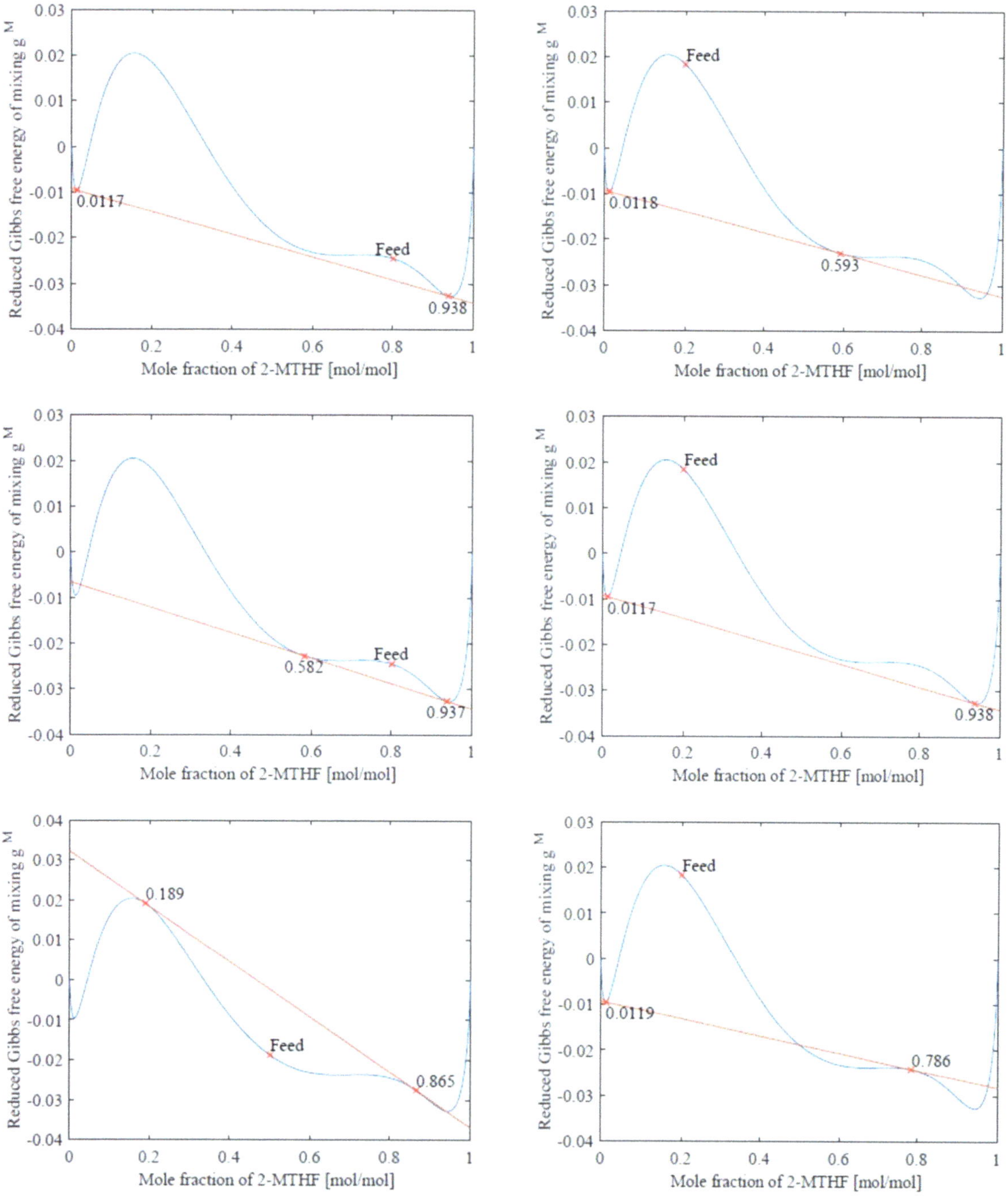

Fig. 5.6.: $g^M(x)$ diagrams for parameter set II. Top left: Secant predicted by DECANTER and FLASH3 (3rd c.). Top right: Secant predicted by FLASH3 and RGIBBS (4th c.). Middle left: Secant predicted by RGIBBS and for Formulation 5.1 by SNOPT (4th c.). Middle right: Secant predicted by DECANTER and by SNOPT for Formulation 5.1 and by BARON for the Formulations 5.1,5.2 and 5.3 (3rd c.). Bottom left: Local maximum predicted by BARON for Formulation 5.2b (4th c.). Bottom right: Saddle predicted by BARON for Formulation 5.2c (4th c.). Aspen Plus DRS predicts $x_{1,1} = 0.786\,\mathrm{mol\,mol^{-1}}$ and $x_{1,2} = 0.0119\,\mathrm{mol\,mol^{-1}}$.

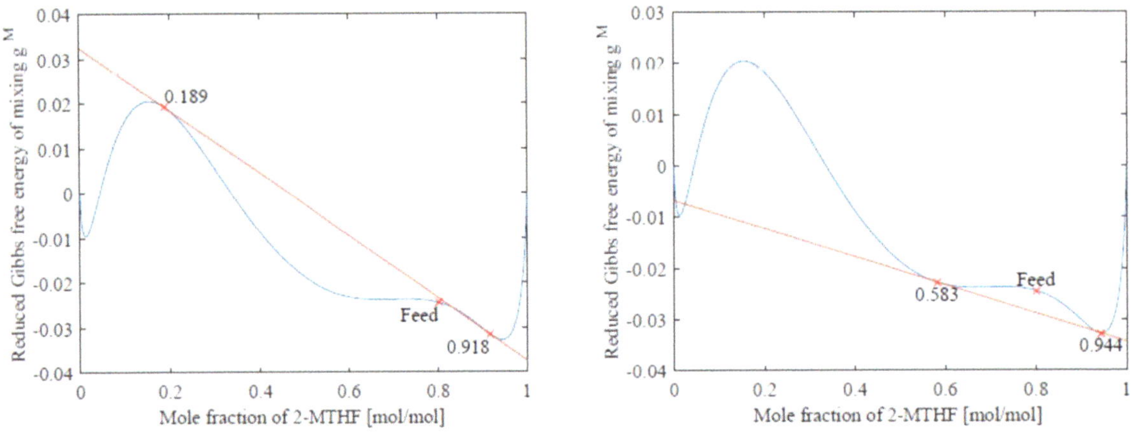

Fig. 5.7.: $g^M(x)$ diagrams for the parameter set II. Left: Saddle (4th c.) predicted by BARON for Formulation 5.2c. Including this saddle and the solutions presented in Figure 5.6, the shown surface of Gibbs free energy has six double tangents, but only one corresponds with the 3rd case. Right: Secant at the compositions predicted by BARON for Formulation 5.3 (4th c.). This solution violates the isoactivity condition. Aspen Plus DRS predicts $x_{1,1} = 0.786\,\mathrm{mol\,mol^{-1}}$ and $x_{1,2} = 0.0119\,\mathrm{mol\,mol^{-1}}$.

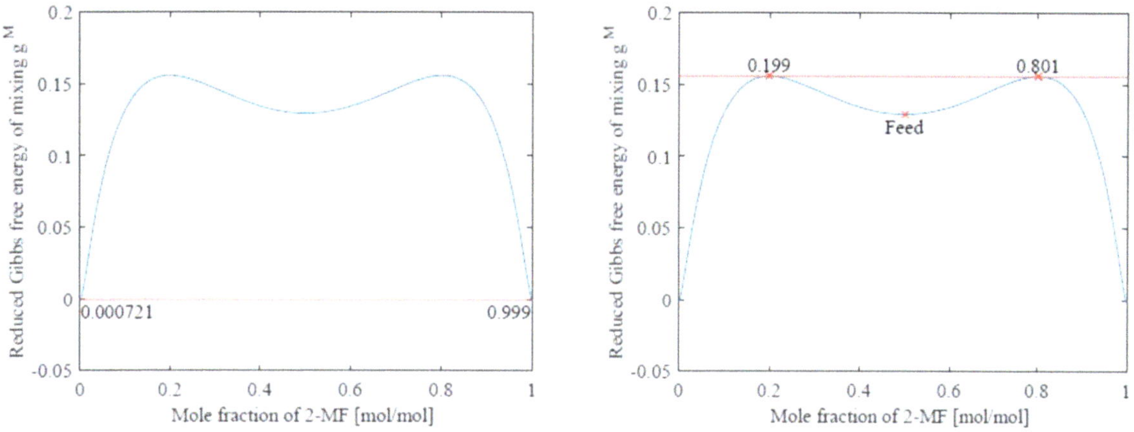

Fig. 5.8.: $g^M(x)$ diagrams for the parameter set III. Left: Secant at compositions predicted by DECANTER and FLASH3 as well as by SNOPT for Formulation 5.1 and by BARON for any of the three Formulations 5.1, 5.2 and 5.3 for both feed compositions in set F_3, and by RGIBBS for the first feed composition (1st case). Right: Global maximum (2nd case) predicted by BARON for Formulation 5.2b. This solution fulfils the isoactivity condition (Equation (2.21)) and could also be returned by other solvers for a formulation that does not impose convexity of the surface of Gibbs free energy at the predicted phase compositions (Equation (5.2a)). At $298.84\,\mathrm{K}$, $x_{1,2} = 0.0007\,\mathrm{mol\,mol^{-1}}$ is measured by Uusi-Kyyny et al. (88) in the aqueous phase.

6. Simulation of combined diffusive and chemical equilibrium

In this chapter, the evaluation of risks is extended to simulation of reactive fluid systems. In these fluid systems, the total composition is not fixed to the feed composition. Thus, the mole balance constraint is relaxed and a system at combined diffusive and chemical equilibrium is predicted. There are two main types of formulations for flash problems with chemical equilibrium (75). On the one hand, stoichiometric formulations use conversions of a set of independent reactions as variables and balances for the species in the fluid system at a given set of conversions as constraints. On the other hand, non-stoichiometric formulations use balances for the elements in the fluid system as constraints.

6.1. Formulations for simulation of combined diffusive and chemical equilibrium

In this section, the formulations that are implemented in GAMS (28) for flash problems of reactive ternary systems are presented. The formulations in this sections are based on Formulation 5.1, but the reduced Gibbs free energy of the complete system is used as objective function because the first term in Equation 2.13 influences the solution of the flash problem for systems with chemical reactions or with different reference state for the phases. Formulation

6.1 represents a non-stoichiometric formulation.

$$\min_{\mathbf{x},\mathbf{n}} \sum_{j \in J} n_j g(\mathbf{x}_j) \tag{6.1a}$$

subject to

$$\sum_{i \in I} x_{i,j} - 1 = 0, \text{ for all } j \in J, \tag{2.9b}$$

$$\sum_{j \in J} \left(n_j \sum_{i \in I} x_{i,j} A_{e,i} \right) = b_e, \text{ for all } e \in E, \tag{6.1b}$$

$$n_j \geqslant 0, \text{ for all } j \in J, \tag{6.2c}$$

$$n_j \leqslant n_{max}, \text{ for all } j \in J, \tag{6.2d}$$

$$x_{i,j} \geqslant x_{min}, \text{ for all } i \in I, j \in J, \tag{5.1c}$$

$$x_{i,j} \leqslant x_{max}, \text{ for all } i \in I, j \in J. \tag{5.1d}$$

Herein, the mole balance is replaced by elemental-abundance constraints (see Equation (6.1b)) and n_j is the amount of moles in phase j and an element of the vector $\mathbf{n}$ and $A_{e,i}$ is the amount of atoms of element e in one molecule of species i and an element of the substance formula matrix $\mathbf{A}$ and b_e is the abundance of element e in mole and an element of the elemental-abundance vector $\mathbf{b}$.

Formulation 6.2 is a stoichiometric formulation for a system with one reaction. To this end, the mole balance from Formulation 5.1 is modified by replacing the constant feed fraction x_i^0 by the fraction of species i in the complete system, which is dependent on the conversion ξ

of the reaction (see Equation (6.2b)).

$$\min_{\mathbf{x},\mathbf{n},\xi} \sum_{j \in J} n_j g\left(\mathbf{x}_j\right) \tag{6.2a}$$

subject to

$$\sum_{i \in I} x_{i,j} - 1 = 0, \text{ for all } j \in J, \tag{2.9b}$$

$$\sum_{j \in J} x_{i,j} n_j = n_i^0 + \nu_i \xi, \text{ for all } i \in I, \tag{6.2b}$$

$$n_j \geqslant 0, \text{ for all } j \in J, \tag{6.2c}$$

$$n_j \leqslant n_{max}, \text{ for all } j \in J, \tag{6.2d}$$

$$x_{i,j} \geqslant x_{min}, \text{ for all } i \in I, j \in J, \tag{5.1c}$$

$$x_{i,j} \leqslant x_{max}, \text{ for all } i \in I, j \in J, \tag{5.1d}$$

$$\xi \geqslant 0, \tag{6.2e}$$

$$\xi \leqslant 1. \tag{6.2f}$$

Herein, n_i^0 is the amount of moles of species i in the complete system at conversion equal to zero and ν_i is the stoichiometric coefficient of species i. The maximal amount of moles in a phase n_{max} must be selected so that a homogeneous phase is feasible for all conversions $\xi \in [0, 1]$. One advantage of Formulation 6.2 is that this problem becomes univariate, if the number of phases is restricted to one.

Of course, many other formulations are possible for the simulation of combined diffusive and chemical equilibrium. For instance, necessary conditions for optimality could be imposed as constraints or the number of variables could be reduced by expressing the variables $x_{i,j}$ and n_j in terms of mole numbers $n_{i,j}$ for all species $i \in I$ and all phases $j \in J$. Formulations including these changes are also implemented in the course of the present work. However, the formulations are not reported here, as no significant differences in the convergence of BARON are observed. The usage of $n_{i,j}$ as variables has the advantage that the bilinear terms in Equations (6.1b) and (6.2b) are avoided so that the feasible set can be guaranteed to be convex.

6.2. Case study: Steam reforming

The aim of this section is to compare results that are reported in the literature that is referenced regarding RGIBBS to the actual results obtained with Aspen Plus in order to find evidence for changes since the implementation of RGIBBS in Aspen Plus. As pointed out in Chapter 5, in the help documentation of Aspen Plus V8.8 (7) Gautam and Seider (29), White and Seider (92) and Schott (71) are referenced regarding RGIBBS. Note that the objective function of the Rand method given by Gautam and Seider (29) in Equation 9 of part one contains bilinear terms and is thus potentially nonconvex. Consequently, several minima may exist. White and Seider (92) give examples for the three approaches that are also available in the RGIBBS unit operation block: *i*) chemical equilibrium, *ii*) temperature approach to chemical equilibrium and *iii*) extents for reactions not taken at equilibrium. We try to reproduce the steam reforming example reported [ibid.] for chemical equilibrium in an ideal gas at $1067\,\mathrm{K}$ and $1\,235\,000\,\mathrm{Pa}$. The involved reactions read

$$\mathrm{CH_4 + H_2O \rightleftharpoons CO + 3\,H_2}, \tag{6.3}$$

$$\mathrm{CO + H_2O \rightleftharpoons CO_2 + H_2}. \tag{6.4}$$

For all species, the model for the ideal gas heat capacities in reads:

$$C_{p,i} = C_1 + C_2 T + C_3 T^2 + C_4 T^3, [C_{p,i}] = \mathrm{J/kmolK}, [T] = \mathrm{K}. \tag{6.5}$$

The same heat capacity coefficients and enthalpies of formation and Gibbs free energies of formation used [ibid.] are employed in the present work. The feed composition and the predicted equilibrium compositions are reported in Table 6.1. As the results show a significant discrepancy, we conclude that either the version of RGIBBS that is currently implemented in Aspen Plus V8.8 (7) is not the same as the version presented by White and Seider (92) or the property models were altered significantly.

6.3. Case study: Hydrolysis of di-sec-butyl ether

As an example for the simulation of combined diffusive and chemical equilibrium the ternary system consisting of sec-butanol (SBA), di-sec-butyl ether (DSBE) and water is selected. The

Table 6.1.: Feed composition and equilibrium compositions predicted by White and Seider (92) and by RGIBBS in the present work for the steam reforming example presented [ibid.]

	Amount of substance in mol		
Species	Feed	Predicitons	
	in both cases	by White and Seider (92)	this work
CH4	15.14	0.875	2.71
H2O	84.07	61.821	63.33
N2	0.59	0.59	0.59
H2	0.2	50.98	45.80
CO	0	6.282	4.12
CO2	0	7.983	8.31

system is subject to the following equilibrium reaction, see e.g. Rossi et al. (67)

$$C_8H_{18}O + H_2O \rightleftharpoons 2\,C_4H_{10}O. \tag{6.6}$$

Kovach and Seider (44; 45) report measurement data of this system and the use of DSBE as an entrainer in an azeotropic distillation for the separation of SBA and water. This process is relevant for the production of 2-butanone by dehydrogenation of SBA. Observation of VLLE on several stages of the distillation column is reported [ibid.]. The LLE measurement data reported by Kovach et al. (45) at 311.15 K and the total composition of the ternary system as a function of the conversion of the hydrolysis reaction of DSBE for a stoichiometric feed are depicted in Figure 6.1.

6.3.1. Methods and materials for the case study

The number of phases of a reactive system in equilibrium is not known a priori. Some authors (48; 24; 29) present phase-splitting algorithms that start with a single phase and increase the number of phases, if no stable solution is found. Nevertheless, in the present work, the optimisation is conducted with the maximum number of phases expected for each phase type, and there is an a posteriori check for trivial solutions which indicate that the number of phases was overestimated. Gibbs' phase rule is used for the maximal number of coexisting phases, see e.g. Pfennig (63)

$$DOF = N + 2 - M - R. \tag{6.7}$$

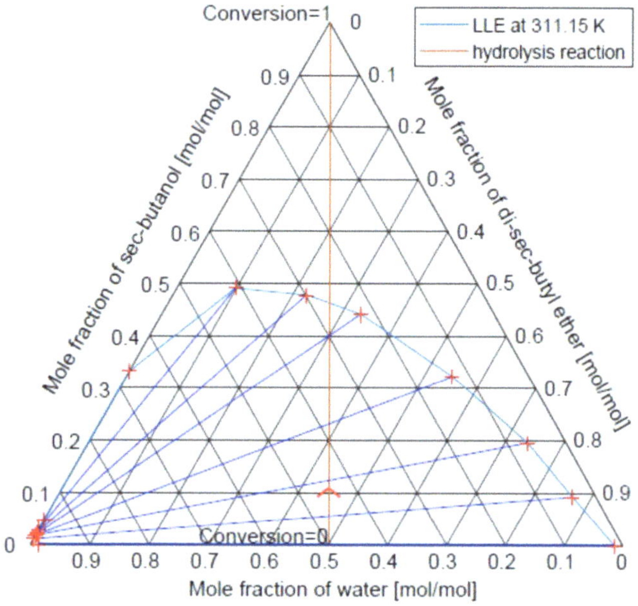

Fig. 6.1.: Ternary diagram of the SBA/DSBE/water system. The LLE data reported by Kovach et al. (45) at 311.15 K (red plus) and the total composition of the ternary system as a function of the conversion of the hydrolysis reaction of DSBE for a stoichiometric feed (red line) are depicted.

Herein, DOF denotes the number of degrees of freedom, M the number of phases, N the number of species and R the number of independent reactions. The maximal number of phases in equilibrium is

$$M_{max} = N + 2 - R. \tag{6.8}$$

In the present work, the maximal number of liquid phases is set to three and the maximal number of vapour phases is set to unity based on Equation (6.8).

In lieu of a regression tool guaranteeing thermodynamic stability for a model of combined diffusive and chemical equilibrium in a ternary system, a regressed model is obtained from the open literature. In principle, a model fitted to measurement data of combined diffusive and chemical equilibrium satisfying thermodynamic stability would be more suitable for the simulation of combined diffusive and chemical equilibrium. For the calculation of the Gibbs free energy of mixing in the liquid phases, the UNIQUAC model (1) with interaction parameters that are regressed to ternary LLE data by Kovach and Seider (45) and the pure-component parameters that are reported [ibid.] is used. Note that measurement of diffusive equilibria for several values of conversion is only possible if chemical equilibrium is kinetically inhibited. However, kinetics of the employed case study are not analysed in the present work. The temperature of the measurement data used in the regression of this model and standard pressure are selected for the solution of the flash problem. The vapour phase is modelled as ideal gas with the ideal Gibbs free energy of mixing calculated from Equation (2.16). The chemical potentials of the pure substances in the vapour phase are computed by the following equation

$$\mu_i^{vapour} = RT \ln\left(\varphi_i^0 (T,p) \frac{p}{p^0}\right) + \int_{T^0}^{T} C_{p,i}^{ig}(T')dT' - T \int_{T^0}^{T} \frac{C_{p,i}^{ig}(T')}{T'} dT' + \mu_i^f + \sum_{e \in E} \mu_e A_{e,i}. \quad (6.9)$$

The fugacity coefficients φ_i^0 of all species in the vapour phase are assumed to be unity. The chemical potentials of the pure substances in the liquid phase are computed by the following equation.

$$\mu_i^{liquid} = F_{P,i} + RT \ln\left(\varphi_i^0 (T, p_i^s(T)) \frac{p_i^s(T)}{p^0}\right) + \int_{T^0}^{T} C_{p,i}^{ig}(T')dT' - T \int_{T^0}^{T} \frac{C_{p,i}^{ig}(T')}{T'} dT' + \mu_i^f + \sum_{e \in E} \mu_e A_{e,i}. \quad (6.10)$$

The Poynting factors $F_{P,i}$ and the fugacity coefficients φ_i^0 of all species at the given temperature and saturation pressure are assumed to be unity. The reference state is defined by the temperature T^0 equal to 298.15 and the pressure p^0 equal to 101325. The standard molar Gibbs free energies of formation μ_i^f at 298.15 are presented in Table 6.2, as taken from the APV88.PURE32 data bank in Aspen Plus V8.8 (7). The following model is used for the heat

Table 6.2.: Standard molar Gibbs free energies of formation μ_i^f at 298.15 obtained from APV88.PURE32 data bank in Aspen Plus V8.8 (7)

	SBA	DSBE	Water
μ_i^f in J/mol	-167000	-115700	-228572

capacities of all three species

$$C_{p,i} = \left(C_1 + C_2 T + C_3 T^2 + C_4 T^3\right) 4.184\,\mathrm{J/cal}. \quad (6.11)$$

The extended Antoine equation that is used for the calculation of the saturation pressures reads

$$p_i^s = \exp\left(A_1 + \frac{A_2}{A_3 + T} + A_4 T + A_5 T^2 + A_6 \ln(T)\right) 101\,325\,\text{Pa/atm}. \tag{6.12}$$

For the extended Antoine equation and the vapour phase heat capacity models, the coefficients reported by Kovach and Seider (44) are used. The indices $i \in \{1, 2, 3\}$ are assigned as follows: SBA (1), DSBE (2), H_2O (3). For the stoichiometric formulation, the mole numbers in the feed n_i^0 at conversion equal to zero are set to 0 mol SBA and 0.5 mol both for DSBE and water. For the non-stoichiometric formulation, the elemental abundance-vector $\mathbf{b}$ is set to $[4, 10, 1]^T$. The substance formula matrix $\mathbf{A}$ reads

$$\begin{pmatrix} 4 & 8 & 0 \\ 10 & 18 & 2 \\ 1 & 1 & 1 \end{pmatrix}. \tag{6.13}$$

This case study is solved in GAMS with the solvers BARON and SNOPT based on the formulations described in section 6.1 and with the unit operation model RGIBBS in Aspen Plus. As reported by White and Seider (92), a non-stoichiometric formulation is used to solve for combined diffusive and chemical equilibrium in the Gibbs algorithm employed in RGIBBS. For all formulations of this case study, the amount of moles in the complete system and n_{max} are both set to unity. Moreover, in GAMS the variables n_j are initialised as unity and the other variables are initialised as described in Section 5.2. Furthermore, the reduced molar Gibbs free energy g is used for plots of the objective function values.

As Kovach et al. (45) do not report whether the UNIQUAC model is regressed guaranteeing thermodynamic stability, the consistency with experimental evidence is evaluated for the binary subsystems, in the present work. The reduced molar Gibbs free energy plots of the binary subsystems are depicted in Figure 6.2. As expected due to the measurements of the ternary LLE by Kovach and Seider (45), no stable vapour phase is predicted for the binary subsystems at 311.15 K and $101\,325$ Pa. While SBA and DSBE are fully miscible, the other two subsystems contain one LLE phase split, each. These results are consistent with the experimental evidence reported [ibid.].

6.3.2. Results

To illustrate the influence of the conversion on the Gibbs free energy of the SBA/DSBE/water system, the Formulation 6.2 is solved for a discretised grid of the conversion and the value of the objective function is then plotted over the conversion in the bottom right diagram of Figure 6.2. The plot indicates that both the function for a homogeneous liquid phase plotted over the conversion and the plot of upper bound values for the objective function obtained with BARON for Formulation 6.2 and discrete conversion are partially convex with respect to the conversion for $\xi \in [0, 1]$. For all tested formulations, BARON does not converge if ξ is variable. Luckily, we have results for discretised ξ. Based on these, we find that the upper bound returned by BARON can be assumed as global optimum within an absolute optimality tolerance of 0.001. The objective function value of this solution is -69.657 mol and the conversion is $0.093\,06$, for details see Table 6.3. It is confirmed that the phases of the predicted LLE fulfil the isoactivity condition. For the same model, RGIBBS predicts a similar solution, see Table 6.3. The

Table 6.3.: Liquid phase compositions for the combined diffusive and chemical equilibrium predicted for the SBA/DSBE/water system at 311.15 K and standard pressure by BARON, RGIBBS and SNOPT, respectively. For the results of RGIBBS and SNOPT the relative deviations from the results of BARON are given.

Phase	BARON	$n_j/\sum_{j\in J} n_j$	$x_{1,j}$	$x_{2,j}$	$x_{3,j}$
Aqueous	value in mol/mol	0.42612	0.00610	0.00051	0.99338
Organic	value in mol/mol	0.57388	0.15763	0.78980	0.05257
	RGIBBS				
Aqueous	value in mol/mol	0.42178	0.00642	0.00051	0.99307
	Δ to BARON	1.0%	5%	2%	0.03%
Organic	value in mol/mol	0.57822	0.16577	0.77913	0.05510
	Δ to BARON	0.8%	5%	1.4%	5%
	SNOPT				
Aqueous	value in mol/mol	0.12059	0.02482	0.00053	0.97465
	Δ to BARON	72%	307%	3%	2%
Organic	value in mol/mol	0.87938	0.43811	0.34774	0.21416
	Δ to BARON	53%	178%	56%	307%

objective function value of this solution is -69.656 mol and the conversion is 0.099. Note that although RGIBBS returned a solution close to the global minimum of the flash problem for this case study, it may also return local minima for reactive flash problems, as it returned local minima for case studies discussed in Chapter 5.

For both Formulation 6.1 and 6.2, SNOPT reports a local minimum with the objective function

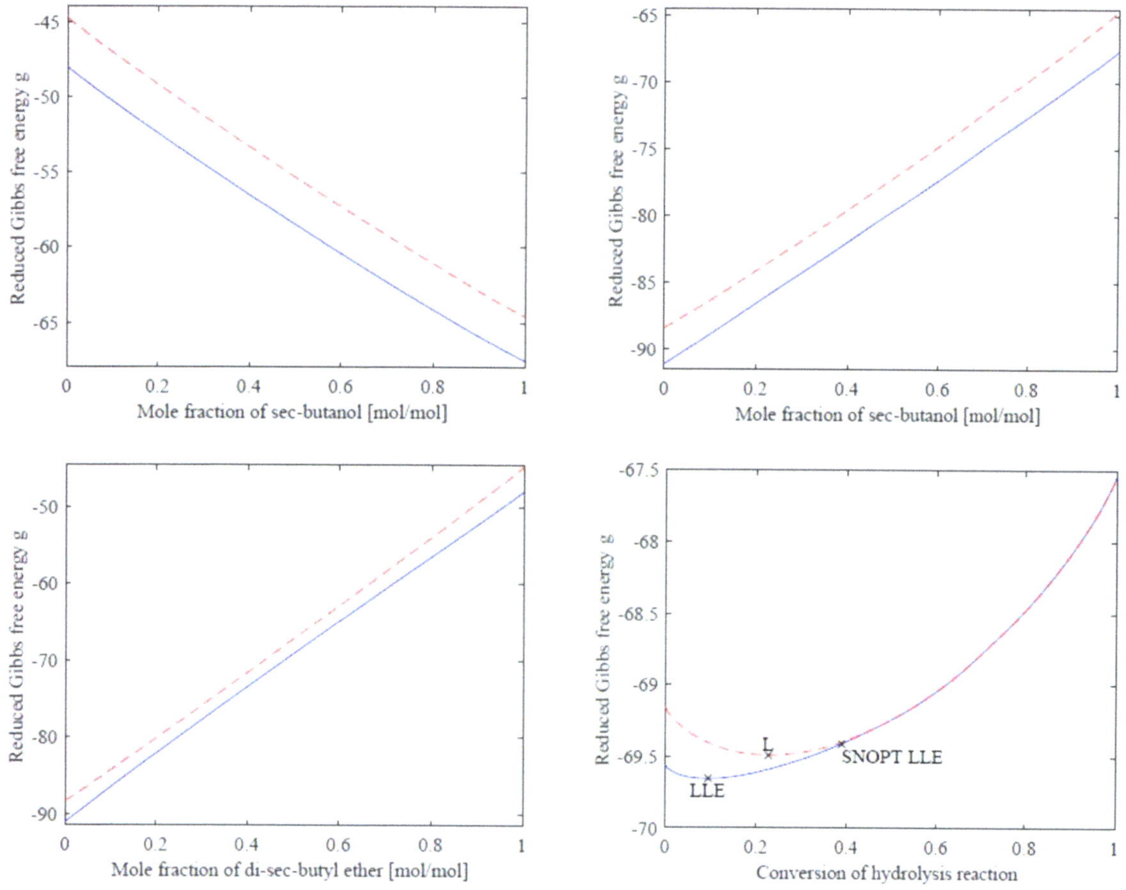

Fig. 6.2.: $g(x)$ diagrams of the binary systems of SBA/DSBE/water and of the hydrolysis of DSBE described by Rossi et al. (67) at $311.15\,K$ and standard pressure. Top left: SBA/DSBE. Top right: SBA/water. Bottom left: DSBE/water. The reduced molar Gibbs free energies for the vapour phase (dashed curves) and for the homogeneous liquid phase (solid curves) are depicted. At these conditions, the binary system SBA/DSBE is correctly modelled to be fully miscible and the binary systems SBA/water and DSBE/water are correctly modelled to contain one LLE phase split. No spurious vapour phase is predicted. Bottom right: The reduced molar Gibbs free energy for a homogeneous liquid phase plotted over the conversion (dashed curve) and the plot of optimal objective function values obtained with BARON for Formulation 6.2 and discrete conversion values (solid curve) are depicted. Both are partially convex with respect to the conversion for $\xi \in [x_{min}, 1-2x_{min}]$. The global minimum of the combined phase and chemical equilibrium problem is marked with "LLE". The minimum of the chemical equilibrium problem restricted to a homogeneous liquid phase is marked with "L". The solution returned by SNOPT for the combined phase and chemical equilibrium is marked by "SNOPT LLE".

value -69.4892 mol and a homogeneous liquid phase with the conversion $0.227\,04$. This is the global minimum for the case restricted to a homogeneous liquid phase however, the objective function value lies significantly above those returned by BARON and RGIBBS and thus, it cannot be the global minimum for combined diffusive and chemical equilibrium. This solution is marked as "L" in the bottom right diagram in Figure 6.2 and also depicted as part of the section through the surface of Gibbs free energy of mixing at the mole fraction $x_{1,j} = 0.227\,04$ mol mol^{-1} in Figure 6.3. The reduced molar Gibbs free energy of mixing g^M is chosen for this plot, as the curvature is the same in the plots of g^M an g however, it is more visible in the plot of g^M because of the higher resolution of the ordinate. As Baker's criterion is not fulfilled, solution "L" cannot be thermodynamically stable. In addition, the Aspen Plus unit operation model REQUIL reports a similar single phase solution with the conversion $0.234\,17$ and the objective function value -69.4890 mol. It is not surprising that REQUIL returns a homogeneous liquid phase, as this model is not able to predict liquid-liquid equilibria, see e.g. Sandler (70).

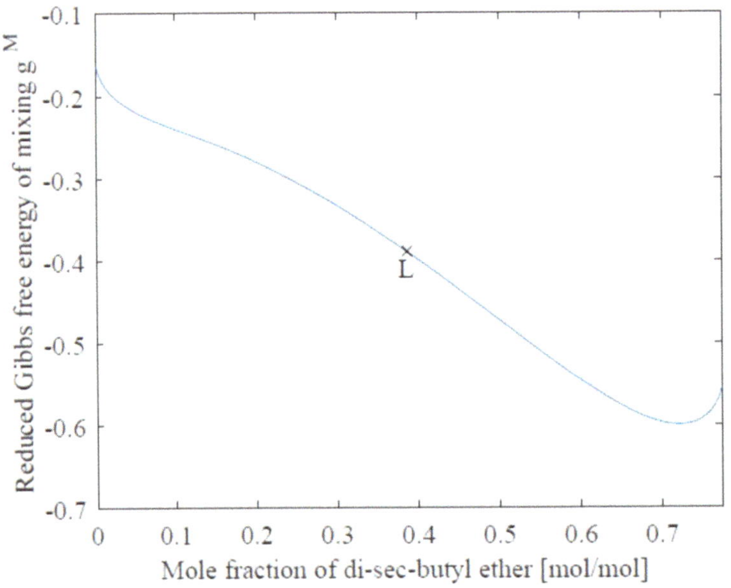

Fig. 6.3.: $g(x)$ diagram of SBA/DSBE/water section at $x_{1,j} = 0.227\,04$ mol mol^{-1}. The solution predicted by SNOPT for both Formulation 6.1 and 6.2 at 311.15 K and standard pressure is marked by "L". However, it is not the global minimum of the flash problem with combined diffusive and chemical equilibrium.

Interestingly, when changing from variables $x_{i,j}$, n_j and ξ to $n_{i,j}$ and the isoactivity condition is imposed for the liquid phases, then SNOPT reports for Formulation 6.2 a LLE as a local minimum with the objective function value $-69.408\,\mathrm{mol}$ and the conversion $0.388\,27$. The phases of the predicted LLE fulfil the isoactivity condition. However, the plot of optimal function values over a discretised grid of conversions in the bottom right diagram in Figure 6.2 does not indicate this solution to be a stationary point. The solution is marked as "SNOPT LLE" in the bottom right diagram in Figure 6.2. The phase compositions for this prediction are printed in Table 6.3. When, on the other hand, Formulation 6.1 is executed with all variables expressed in terms of mole numbers $n_{i,j}$ and isoactivity condition imposed for the liquid phases, then SNOPT terminates the optimisation due to difficulties with a singular basis and reports a feasible solution that indicates a VLE. This solution is not assumed to be a local minimum as it violates the isopotential condition (see Equation (2.18)). On the one hand, the differences between the stoichiometric and the non-stoichiometric formulations regarding convergence to the global minimum are not significant. On the other hand, the global solver BARON (83) seems to have significant advantages compared to local solvers like SNOPT (31).

Note that the kinetics of the employed case study are not analysed in the present work. Although kinetics are relevant in the modelling of reactive systems just as they are for diffusive mass transfer, the scope of the present work is restricted to prediction of equilibrium states.

After all, agreement between RGIBBS and both Formulation 6.1 and 6.2 solved with BARON is observed. For the employed case study, no Formulation is superior over the other in terms of convergence to the global minimum. Nevertheless, for both formulations, there remain unresolved issues with SNOPT.

7. Conclusion and outlook

The purpose of the present work is *i)* to obtain NRTL parameter sets representing good fits to the aforementioned TMFB case studies and guaranteeing thermodynamic stability, as well as the correct numbers of phase splits and phases per phase split and *ii)* to discuss formulations for parameter estimation for thermodynamic models of diffusive equilibria and formulations for the simulation of diffusive equilibrium and combined diffusive and chemical equilibrium based on regressed thermodynamic models. The main objective of the present work is to demonstrate how inadequate regression techniques can affect results in computer-aided process engineering, and, if possible, to reverse-engineer what formulations are used by Aspen Plus, including unit operation models and the Aspen Plus Data Regression System.

Summing up the results of Chapter 3, five types of formulations for parameter estimation for liquid-liquid equilibria and the kind of guarantees they can provide are discussed. The first type of formulation does not guarantee to fulfil the well-known isoactivity condition, which can be derived from 1st-order KKT conditions of the flash problem. The second type of formulation guarantees the predictions to be stationary points of the respective flash problems by employing 1st-order necessary conditions like the well-known isoactivity condition, while the third type guarantees the predictions to be local minima of the respective flash problems by employing 2nd-order sufficient KKT conditions. In addition to 1st-order or 2nd-order KKT conditions, type four employs further conditions to reduce the risk to predict the wrong numbers of phase splits and phases per phase split. On top of that, the fifth type that is presented employs criteria that are both sufficient and necessary for stability. This includes Baker's criterion (8), as well as criteria to guarantee the correct numbers of phases and phase splits.

In Chapter 4, a formulation of the latter type is used through BOARPET (58) to regress NRTL parameter sets for five challenging binary LLE systems relevant to the cluster of excellence TMFB. In particular, to the author's best knowledge, this work is the first to report a model for the LLE of the 2,5-DMTHF/water system. In addition, parameter sets that allow modelling of the LLE of 2,5-DMF/water, 2-MTHF/water and butanone/water are regressed. The same measurement data used in BOARPET, are used for regression in Aspen Plus V8.8 (7), to compare both regression methods. It is found that for butanone/water the regression results of DRS violate thermodynamic criteria. First, the resulting model contains a spurious second phase split. In addition, at a certain temperature, the phase compositions that are predicted by Aspen Plus DRS based on this model do not fulfil the 1st-order necessary KKT conditions of the binary LLE flash problem. The user may not be aware of this, as the status is "Results Available".

In Chapter 5, the risks of inadequate regression techniques in subsequent process simulation are assessed. Three different unit operation models in Aspen Plus are tested. In order to reverse-engineer the methods that are implemented in these models, three different solvers are applied to several case studies based on three different formulations of the LLE flash simulation in GAMS (28). The respective parameter sets of the employed G^E model are selected from the first part, either because these are found to result in violation of thermodynamics, or due to the particular shape of the implied Gibbs free energy curve. While two of the formulations implemented in GAMS are discussed extensively in literature as phase splitting methods, Formulation 5.3 is usually used as a stability test, see e.g. McDonald and Floudas (52).

Four cases are discriminated in the evaluation of risks for process simulation. In the 1st case, the regressed model satisfies thermodynamic stability as well as correct numbers of phase splits and phases per phase split, and the flash problem is solved to global optimality. Only in this case the risk of erroneous prediction of diffusive equilibrium is avoided, e.g., regarding process simulation. In the 2nd case, the regressed model satisfies the mentioned conditions, but the flash solver returns a solution that is significantly different from the globally optimal solution. In the remaining cases, the regressed model violates these thermodynamic constraints and the flash solver returns the global minimum (3rd case) or a different solution (4th case). It is observed that the Aspen Plus unit operation models DECANTER, FLASH3 and RGIBBS return solutions of the flash problem that are significantly different from the global minimum. Hence, when employing these unit operation models obtaining the 1st case cannot be guaranteed. Furthermore, the error in the prediction of diffusive equilibrium may propagate in unit operation models that contain more than one flash simulation. Conclusively, the author recommends employing global solvers, e.g. BARON (83), in modelling environments like GAMS (28) or AMPL (2) for the simulation of diffusive equilibria.

In Chapter 6, stoichiometric and non-stoichiometric formulations from literature for the simulation of combined diffusive and chemical equilibrium are compared using local and global solvers through GAMS, using the case study SBA/DSBE/water. As agreement between the results of RGIBBS in Aspen Plus and BARON in GAMS is good, to date, no conclusions can be drawn on the risks, potentially resulting from Aspen Plus unit operation models for reactive mixtures.

As a result of the present work, the author would like to inform the reader about the risk of obtaining qualitatively wrong predictions of diffusive equilibria, if inadequate regression techniques and/or unit operation models are employed.

The comparison of BOARPET to other regression tools like the Aspen Plus Data Regression System must, to date, be restricted to diffusive equilibria of binary fluid systems. As future work, BOARPET should be extended to multicomponent systems, which is, in principle, allowed by the underlying method but not implemented, yet. This extension would pave the way to further challenging case studies of diffusive equilibrium for the comparison with other regression tools. While the unit operation model RGIBBS is observed to return a local (as opposed to the global) minimum in the simulation of diffusive equilibrium, assessment of risks resulting from simulation of combined diffusive and chemical equilibrium with RGIBBS remains an open issue. In order to extend the evaluation of risks regarding predictions of combined diffusive and chemical equilibrium that are employed in process simulation, more challenging reactive fluid systems should be studied. For instance, such case studies are listed by Bonilla-Petriciolet et al. (13) together with references for the corresponding thermodynamic models. Lastly, BOARPET should be extended to reactive fluid systems so that models for further case studies of combined diffusive and chemical equilibrium can be regressed guaranteeing thermodynamic stability, as well as the correct numbers of phase splits and phases per phase split.

In general, it is crucial to find a model that is fit for purpose. Simpler models that are not based on physical approaches can avoid the challenges of bi-level optimisation. However, they cannot deliver the same level of insight that the models discussed in the present work can offer. For models with multi-level optimisation, global optimisation is crucial to deliver meaningful results. As common commercial software for optimisation does not visualise these errors, the failure to achieve global optimisation can easily be overlooked by practitioners during numerical process simulation. Future work could focus on the design of a user interface that makes it easier to identify such failures.

Bibliography

[1] D. S. Abrams and J. M. Prausnitz. Statistical thermodynamics of liquid mixtures: A new expression for the excess Gibbs energy of partly or completely miscible systems. *AIChE Journal*, 21(1):116–128, 1975.

[2] AMPL Optimization Inc. AMPL Release 20131012, 2013.

[3] V. P. Aneja. Organic Compounds in Cloud Water and their Deposition at a Remote Continental Site. *Air & Waste*, 43(9):1239–1244, 1993.

[4] A. Arce, A. Blanco, P. Souza, and I. Vidal. Liquid-Liquid Equilibria of the Ternary Mixtures Water + Propanoic Acid + Methyl Ethyl Ketone and Water + Propanoic Acid + Methyl Propyl Ketone. *Journal of Chemical & Engineering Data*, 40(1):225–229, 1995.

[5] Artelys, France. Artelys Knitro Release 10.2.0, 2016.

[6] Aspen Technology, Inc. aspenONE Version 8.6, 2014.

[7] Aspen Technology, Inc. aspenONE Version 8.8, 2015.

[8] L. E. Baker, A. C. Pierce, and K. D. Luks. Gibbs Energy Analysis of Phase Equilibria. *Society of Petroleum Engineers Journal*, pages 731–742, 1982.

[9] G. M. Bennett and W. G. Philip. The influence of structure on the solubilities of ethers. Part II. Some cyclic ethers. *Journal of the Chemical Society*, pages 1937–1942, 1928.

[10] L. C. Blumenthal, C. M. Jens, J. Ulbrich, F. Schwering, V. Langrehr, T. Turek, U. Kunz, K. Leonhard, and R. Palkovits. Systematic Identification of Solvents Optimal for the Extraction of 5-Hydroxymethylfurfural from Aqueous Reactive Solutions. *ACS Sustainable Chemistry & Engineering*, 4(1):228–235, 2016.

[11] A. Boldyrev, V. Komarov, and V. Krichevtsov. Equilibriums in systems formed by components of the catalytic reactions of furfural. *Zh. Prikl. Khim.(Leningrad)*, 46:2336–2339, 1973.

[12] G. M. Bollas, P. I. Barton, and A. Mitsos. Bilevel optimization formulation for parameter estimation in vapor-liquid(-liquid) phase equilibrium problems. *Chemical Engineering Science*, 64:1768–1783, 2009.

[13] A. Bonilla-Petriciolet, M. d. R. Moreno-Virgen, and J. J. Soto-Bernal. Global Gibbs free energy minimization in reactive systems via harmony search. *International Journal of Chemical Reactor Engineering*, 10:1–23, 2012.

[14] J. F. Boston. *Inside-Out Algorithms for Multicomponent Separation Process Calculations*, chapter 6, pages 135–151. ACS symposium series. American Chemical Society, 1980.

[15] J. F. Boston and P. M. Mathias. *Phase Equilibria in a Third-Generation Process Simulator*, pages 822–848. EFCE publication series. American Institute of Chemical Engineers and European Federation of Chemical Engineering and Dechema, 1980.

[16] S. Brandt. *Datenanalyse: Mit statistischen Methoden und Computerprogrammen*. Spektrum Akademischer Verlag, Heidelberg, 4th edition, 1999.

[17] S. Cabani, G. Conti, and L. Lepori. Thermodynamic Study on Aqueous Dilute Solutions of Organic Compounds. Part 2. - Cyclic Ethers. *Transactions of the Faraday Society*, 67:1943–1950, 1971.

[18] A. N. Campbell, E. M. Kartzmark, and W. E. Falconer. The system: nicotine - methylethyl ketone - water. *Canadian Journal of Chemistry*, 36(11):1475–1486, 1958.

[19] Chemstations, Inc. CHEMCAD 5.2.0. (Chemical Process Simulation), 1998-2002.

[20] J. L. da Silva Jr. and M. Aznar. Thermophysical properties of 2,5-dimethylfuran and liquid-liquid equilibria of ternary systems water + 2,5-dimethylfuran + alcohols (1-butanol or 2-butanol or 1-hexanol). *Fuel*, 136:316 – 325, 2014.

[21] D. S. Dechambre. *Efficient Measurement of Liquid-Liquid Equilibria Using Automation and Optimal Experimental Design*. PhD thesis, Lehrstuhl für Technische Thermodynamik (LTT), RWTH Aachen University, Germany, 2016.

[22] D. S. Dechambre, J. Thien, and A. Bardow. When 2nd Generation Biofuel meets Water - The Water Solubility and Phase Stability Issue. *Manuscript submitted for publication*, 2016.

[23] DECHEMA, Gesellschaft für Chemische Technik und Biotechnologie e.V., Germany. Data Preparation Package (DPP), 2014.

[24] F. Denes, P. Lang, and M. Lang-Lazi. Liquid-Liquid-Liquid Equilibrium Calculations. In *Institution of Chemical Engineers Symposium Series*, volume 152, pages 877–890. Institution of Chemical Engineers, 2006.

[25] M. Eidinoff. Distillation of water. In G. Murphy, H. Urey, and I. Kirshenbaum, editors, *Production of Heavy Water. Part II: Laboratory Studies for Separation Processes*, chapter 7, pages 120–144. McGraw-Hill Book Co., Inc., New York, 1955.

[26] P. Englezos and N. Kalogerakis. *Applied parameter estimation for chemical engineers*. Marcel Dekker, Inc., New York, 2001.

[27] I. Ferino, B. Marongiu, V. Solinas, and S. Torrazza. Thermodynamic properties of aqueous non-electrolyte mixtures. II. Excess enthalpies and liquid-liquid equilibrium of 2-alkanone + water mixtures. *Thermochimica Acta*, 70(1):149 – 161, 1983.

[28] GAMS Development Corporation. General Algebraic Modeling System (GAMS) Release 24.8.4, 2017.

[29] R. Gautam and W. D. Seider. Computation of phase and chemical equilibrium: Parts I to III. *AIChE Journal*, 25(6):991–1015, 1979.

[30] R. C. Ghildiyal, S. M. Nanoti, N. N. Kulsrestha, B. S. Rawat, Y. Kumar, and R. Krishna. Liquid-liquid equilibrium in the toluene - methyl ethyl ketone - water system. *Fluid Phase Equilibria*, 50(3):339 – 346, 1989.

[31] P. E. Gill, W. Murray, and M. A. Saunders. SNOPT: An SQP algorithm for large-scale constrained optimization. *SIAM Rev.*, 47:99–131, 2005.

[32] P. E. Gill, W. Murray, M. A. Saunders, and E. Wong. User's Guide for SNOPT 7.6: Software for Large-Scale Nonlinear Programming. Center for Computational Mathematics Report CCoM 17-1, Department of Mathematics, University of California, San Diego, La Jolla, CA, 2017.

[33] M. Glass, M. Aigner, J. Viell, A. Jupke, and A. Mitsos. Liquid-liquid equilibrium of 2-methyltetrahydrofuran/water over wide temperature range: Measurements and rigorous regression. *Fluid Phase Equilibria*, 433:212 – 225, 2017.

[34] M. Glass, J.-R. Bruch, and A. Mitsos. Modeling Miscibility Gaps in Water/Platform Molecule Binary Mixtures. Technical report, AVT Process Systems Engineering (AVT.SVT), RWTH Aachen University, Germany, May 2017.

[35] M. Glass, J.-R. Bruch, J. Viell, and A. Mitsos. Modeling Miscibility Gaps in Water/Platform Molecule Binary Mixtures, presented as a poster at the 5th International Conference of the Cluster of Excellence "Tailor-Made Fuels from Biomass", June 20 to 22, 2017 in Aachen, Germany.

[36] M. Glass, H. Djelassi, and A. Mitsos. Bilevel Programming in Parameter Estimation of Cubic Equations of State. Technical report, AVT Process Systems Engineering (AVT.SVT), RWTH Aachen University, Germany, 2017.

[37] M. Glass and A. Mitsos. Notes on Comparison of BOARPET (DPP) to Gauss-Newton (DPP). Technical report, AVT Process Systems Engineering (AVT.SVT), RWTH Aachen University, Germany, June 2017.

[38] M. Glass and A. Mitsos. Parameter Estimation for Phase Equilibrium Problems via Bilevel Programs, presented as a talk at "ProcessNet AA-MPO Meeting", May 12, 2017 in Leverkusen, Germany.

[39] S. Gouli, E. Lois, and S. Stournas. Effects of Some Oxygenated Substitutes on Gasoline Properties, Spark Ignition Engine Performance, and Emissions. *Energy & Fuels*, 12(5):918–924, 1998.

[40] M. Hechinger. *Model-based Identification of Promising Biofuel Candidates for Spark-ignited Engines*, volume 940 of *Fortschritt-Berichte VDI. Reihe 3, Verkehrstechnik*. VDI Verlag, Düsseldorf, 2014.

[41] F. Hoppe, U. Burke, M. Thewes, K. A. Heufer, F. Kremer, and S. Pischinger. Tailor-Made Fuels from Biomass: Potentials of 2-butanone and 2-methylfuran in direct injection spark ignition engines. *Fuel*, 167:106 – 117, 2016.

[42] B. Houben, S. Schuback, J. Wurlitzer, Dombrowski, and J. Klankermayer. HPLC-Methodenentwicklung zur Quantifizierung von Monosacchariden im Zweiphasensystem. Projektarbeit, Institut für Technische und Makromolekulare Chemie (ITMC), RWTH Aachen University, Germany, in cooperation with Berufskolleg Simmerath/Stolberg der Städteregion Aachen, 2012.

[43] A. Y. Karmilchik and L. S. Efimova. Study of mutual solubility in a three component system 5-methylfurfural - sylvane - water. *Latvijas PSR Zinatnu Akademijas Vestis, Kimijas Serija*, 3:282–286, 1970.

[44] J. W. Kovach III and W. D. Seider. Heterogeneous azeotropic distillation: Experimental and simulation results. *AIChE Journal*, 33(8):1300–1314, 1987.

[45] J. W. Kovach III and W. D. Seider. Vapor-liquid and liquid-liquid equilibria for the system sec-butyl-alcohol-di-sec-butyl ether-water. *Journal of Chemical & Engineering Data*, 33(1):16–20, 1988.

[46] E. Lladosa, J. B. Montón, J. de la Torre, and N. F. Martínez. Liquid-Liquid and Vapor-Liquid-Liquid Equilibrium of the 2-Butanone + 2-Butanol + Water System. *Journal of Chemical & Engineering Data*, 56(5):1755–1761, 2011.

[47] K. Lucas. *Molecular models for fluids*. Cambridge University Press, New York, 2007.

[48] A. Lucia, L. Padmanabhan, and S. Venkataraman. Multiphase equilibrium flash calculations. *Computers & Chemical Engineering*, 24(12):2557–2569, 2000.

[49] A. Lucia and J. Xu. Chemical process optimization using Newton-like methods. *Computers & Chemical Engineering*, 14(2):119–138, 1990.

[50] M. Männistö, J.-P. Pokki, L. Fournis, and V. Alopaeus. Ternary and binary LLE measurements for solvent (2-methyltetrahydrofuran and cyclopentyl methyl ether) + furfural + water between 298 and 343 K. *The Journal of Chemical Thermodynamics*, 110:127 – 136, 2017.

[51] A. Marcilla, M. d. M. Olaya, and M. D. Serrano. Comments on Liquid-Liquid Equilibrium Data Regression. *Journal of Chemical & Engineering Data*, 52(6):2538–2541, 2007.

[52] C. M. McDonald and C. A. Floudas. Global optimization for the phase stability problem. *AIChE Journal*, 41(7):1798–1814, 1995.

[53] V. A. Merchan, G. Tolksdorf, R. Kraus, and G. Wozny. Extending Documentation-Based Models towards an Efficient Integration into Commercial Process Simulators. *Chemie Ingenieur Technik*, 86(7):1117–1129, 2014.

[54] M. L. Michelsen. The isothermal flash problem. Part I. Stability. *Fluid Phase Equilibria*, 9(1):1 – 19, 1982.

[55] A. Mitsos and P. I. Barton. A Dual Extremum Principle in Thermodynamics. *AIChE Journal*, 53:2131–2147, 2007.

[56] A. Mitsos, G. M. Bollas, and P. I. Barton. Bilevel optimization formulation for parameter estimation in liquid-liquid phase equilibrium problems. *Chemical Engineering Science*, 64:548–559, 2009.

[57] A. Mitsos, G. M. Bollas, and P. I. Barton. Model and parameter identification in phase equilibria. *19th European Symposium on Computer Aided Process Engineering - ESCAPE-19, Cracow, Poland*, 2009.

[58] A. Mitsos, T. Hoffmann, and M. Glass. Bilevel Optimization Algorithm for Rigorous and Robust Parameter Estimation in Thermodynamics (BOARPET) v1.3. March 2017.

[59] J. A. Nelder and R. Mead. A simplex method for function minimization. *The Computer Journal*, 7(4):308–313, 1965.

[60] J. Nocedal and S. J. Wright. *Numerical optimization*. Springer series in operations research and financial engineering. Springer, Berlin and New York, 2nd edition, 2006.

[61] J. P. Novák, J. Matouš, J. Pick, E. Hála, E. Erdös, and M. Štulíková. *Liquid-liquid equilibria*, volume 7 of *Studies in modern thermodynamics*. Elsevier, Amsterdam, 1987.

[62] K. Ochi, M. Tada, and K. Kojima. Measurement and correlation of liquid-liquid equilibria up to critical solution temperature. *Fluid Phase Equilibria*, 56:341 – 359, 1990.

[63] A. Pfennig. *Thermodynamik der Gemische*. Springer, Berlin and New York, 2004.

[64] M. S. Qureshi, P. Vrbka, and V. Dohnal. Thermodynamic properties of five biofuel-relevant compounds at infinite dilution in water. *Fuel*, 191:518 – 527, 2017.

[65] H. Renon and J. M. Prausnitz. Local compositions in thermodynamic excess functions for liquid mixtures. *AIChE Journal*, 14(1):135–144, 1968.

[66] Y. Roman-Leshkov, C. J. Barrett, Z. Y. Liu, and J. A. Dumesic. Production of dimethylfuran for liquid fuels from biomass-derived carbohydrates. *Nature*, 447(7147):982–985, 2007.

[67] C. Rossi, L. Cardozo-Filho, and R. Guirardello. Gibbs free energy minimization for the calculation of chemical and phase equilibrium using linear programming. *Fluid Phase Equilibria*, 278(1):117 – 128, 2009.

[68] T. W. Rudolph and J. J. Thomas. NOx, NMHC and CO emissions from biomass derived gasoline extenders. *Biomass*, 16(1):33 – 49, 1988.

[69] F. Ruiz, M. I. Galan, D. Prats, and A. M. Ancheta. Temperature influence on the ternary system 1-butanol-butanone-water. *Journal of Chemical & Engineering Data*, 29(2):143–146, 1984.

[70] S. I. Sandler. *Using Aspen Plus in Thermodynamics Instructions: A Step by Step Guide*. John Wiley & Sons, Inc., Hoboken, NJ, USA, 2015.

[71] G. L. Schott. Computation of Restricted Equilibria by General Methods. *The Journal of Chemical Physics*, 40(7):2065–2066, 1964.

[72] A. E. Skrzec and N. F. Murphy. Liquid-Liquid Equilibria. Acetic Acid in Water, with 1-Butanol, Methyl Ethyl Ketone, Furfural, Cyclohexanol, and Nitromethane. *Industrial & Engineering Chemistry*, 46(11):2245–2247, 1954.

[73] A. S. Smith and J. F. LaBonte. Dehydration of Aqueous Methyl Ethyl Ketone with 2-Methyl Furan. *Industrial & Engineering Chemistry*, 44(11):2740–2743, 1952.

[74] J. V. Smith, R. W. Missen, and W. R. Smith. General optimality criteria for multiphase multireaction chemical equilibrium. *AIChE Journal*, 39(4):707–710, 1993.

[75] W. R. Smith. The Computation of Chemical Equilibria in Complex Systems. *Industrial & Engineering Chemistry Fundamentals*, 19(1):1–10, 1980.

[76] J. M. Sørensen and W. Arlt. *Liquid-liquid equilibrium data collection: Binary systems: Tables, diagrams and model parameters*, volume 5, part 1 of *Chemistry data series*. DECHEMA Deutsche Gesellschaft für Chemisches Apparatewesen, Frankfurt/Main, 1979.

[77] J. M. Sørensen and W. Arlt. *Liquid-liquid equilibrium data collection: Ternary and quaternary systems*, volume 5, part 3 of *Chemistry data series*. DECHEMA Deutsche Gesellschaft für Chemisches Apparatewesen, Frankfurt/Main, 1980.

[78] J. M. Sørensen and W. Arlt. *Liquid-liquid equilibrium data collection: Ternary systems*, volume 5, part 2 of *Chemistry data series*. DECHEMA Deutsche Gesellschaft für Chemisches Apparatewesen, Frankfurt/Main, 1980.

[79] J. M. Sørensen, T. Magnussen, P. Rasmussen, and A. Fredenslund. Liquid-liquid equilibrium data: Their retrieval, correlation and prediction Part II: Correlation. *Fluid Phase Equilibria*, 3(1):47 – 82, 1979.

[80] P. Stephan, K. Schaber, K. Stephan, and F. Mayinger. *Thermodynamik: Grundlagen und technische Anwendungen – Band 2: Mehrstoffsysteme und chemische Reaktionen*, volume 2. Springer, Berlin and New York, 15th edition, 2010.

[81] R. M. Stephenson. Mutual solubilities: water-ketones, water-ethers, and water-gasoline-alcohols. *Journal of Chemical Engineering Data*, 37(1):80–95, 1992.

[82] L. Stragevitch. *Liquid-liquid equilibrium in non-electrolyte mixtures (in Portuguese)*. PhD thesis, School of Chemical Engineering, University of Campinas, 1997.

[83] M. Tawarmalani and N. V. Sahinidis. A polyhedral branch-and-cut approach to global optimization. *Mathematical Programming*, 103:225–249, 2005.

[84] J. Taylor. *Introduction to Error Analysis, the Study of Uncertainties in Physical Measurements*. University Science Books, 2nd edition, Aug. 1997.

[85] S.-J. Tenne, J. Kinzel, M. Arlt, F. Sibilla, M. Bocola, and U. Schwaneberg. 2-Methyltetrahydrofuran and cyclopentylmethylether: Two green solvents for efficient purification of membrane proteins like FhuA. *Journal of Chromatography B*, 937:13 – 17, 2013.

[86] J. W. Tester and M. Modell. *Thermodynamics and its applications*. Prentice-Hall international series in the physical and chemical engineering sciences. Prentice Hall PTR, Upper Saddle River, N.J., 3rd edition, 1997.

[87] U. Westhaus from DECHEMA Gesellschaft für Chemische Technik und Biotechnologie e.V. Personal communication with Moll Glass, 2017/02/01.

[88] P. Uusi-Kyyny, S. Ionita, M. S. Qureshi, V. Alopaeus, and D. Richon. Design of Equilibrium Cells for Phase Equilibria and PVT Measurements in Large Ranges of Temperatures and Pressures. I. Vapor-Liquid-Liquid Equilibria. *Journal of Chemical & Engineering Data*, 61(8):2700–2711, 2016.

[89] A. van den Bos. *Parameter estimation for scientists and engineers*. John Wiley & Sons, Inc., Hoboken, NJ, USA, 2007.

[90] S. M. Walas. *Phase Equilibria in Chemical Engineering*. Butterworth Publishers, Boston, MA, USA, 1985.

[91] C. Wang, Y. Chen, K. Cheng, and C. Wen. Phase equilibria of the ternary systems of (water + diethyl carbonate) with acetone, or 2-butanone at four temperatures. *The Journal of Chemical Thermodynamics*, 102:147 – 154, 2016.

[92] C. W. White III and W. D. Seider. Computation of phase and chemical equilibrium, part IV: Approach to chemical equilibrium. *AIChE Journal*, 27(3):466–471, 1981.

[93] F. Wieland, A. Neff, A. N. Gloess, L. Poisson, S. Atlan, D. Larrain, D. Prêtre, I. Blank, and C. Yeretzian. Temperature dependence of Henry's law constants: An automated, high-throughput gas stripping cell design coupled to PTR-ToF-MS. *International Journal of Mass Spectrometry*, 387:69 – 77, 2015.

[94] W. Yang and A. Sen. One-Step Catalytic Transformation of Carbohydrates and Cellulosic Biomass to 2,5-Dimethyltetrahydrofuran for Liquid Fuels. *ChemSusChem*, 3(5):597–603, 2010.

A. Fundamentals of parameter estimation

The optimisation problem arising when an existing model shall be fitted to measurement data is usually named parameter estimation in an engineering context or regression analysis in a statistical context (26). In the present work, the terms regression and estimation are used synonymously. The objective functions employed in problems of this kind measure the departure between a measured value x^{exp} of a regressed variable and the predicted value x^{est} of this regressed variable for several experiments. The difference between these values for experiment k is the residual

$$r_k = x_k^{exp} - x_k^{est}. \tag{A.1}$$

If several variables are regressed for a single experiment, the residual $\mathbf{r}_k$ is a vector. The vector of predicitions $\mathbf{x}^{exp}$ depends on a vector of regressed parameters $\mathbf{p}$ and a vector of fixed variables $\mathbf{t}$. These fixed variables are also measured and not necessarily constant for all experiments. However, they are not regressed and their uncertainty is assumed to be negligible compared to the uncertainty of the regressed variables. Regarding parameter estimation in engineering practice, objective functions that are based on the Least Square function or the Maximum Likelihood function are most important (89). The general Least Square function reads

$$\sum_{k \in K} \mathbf{r}_k^T \mathbf{w}_k \mathbf{r}_k, \tag{A.2}$$

where K is the index set of experiments and $\mathbf{w}_k$ is the weighting matrix for experiment k. Moreover, the Least Squares objective function utilised in BOARPET reads

$$LS = \sum_{k=1}^{L} \sum_{j=1}^{2} w_{j,k} \left(x_{j,k}^{\mathrm{exp}} - x_{j,k}^{\mathrm{est}} \right)^2, \tag{A.3}$$

where k denotes the experiments, L the total number of experiments, K the index set of experiments $K = \{1, 2, \ldots L\}$, j the phases, J the index set of phases $J = \{1, 2\}$, $x_{j,k}$ the mole fraction of butanone in phase j of experiment k, and $w_{j,k}$ the weight of phase j of experiment k. Similarly, the general Maximum Likelihood function reads

$$\left(\prod_{k\in K} \det\left(\mathbf{w}_k\right)\right)^{-\frac{1}{2}} \exp\left(-\frac{1}{2}\sum_{k\in K} \mathbf{r}_k^T \mathbf{c}_k^{-1} \mathbf{r}_k\right), \tag{A.4}$$

where $\mathbf{c}_k$ is the covariance matrix of $\mathbf{r}_k$. This is the joint probability density function for all regressed variables, if the model is adequate and the errors in the different experiments are independent and distributed normally, see e.g. Englezos and Kalogerakis (26). If only one regressed variable is measured in each experiment the Maximum Likelihood function can be expressed as follows, see e.g. Brandt (16):

$$\prod_{k\in K}\left(\left(2\pi\sigma_k^2\right)^{-\frac{1}{2}} \exp\left(\frac{-r_k^2}{2\sigma_k^2}\right)\right), \tag{A.5}$$

where σ_k is the standard deviation of the normal distribution of the regressed variable of experiment k. The Maximum Likelihood objective function that is implemented in the DRS Regression in Aspen Plus reads

$$\sum_{n=1}^{NDG} w_n \sum_{m=1}^{NP_n}\left[\left(\frac{T_m^{\text{exp}}-T_m^{\text{est}}}{\sigma_m^T}\right)^2+\left(\frac{p_m^{\text{exp}}-p_m^{\text{est}}}{\sigma_m^P}\right)^2+\sum_{i=1}^{N-1}\left(\frac{x_{m,i}^{\text{exp}}-x_{m,i}^{\text{est}}}{\sigma_{m,i}^x}\right)^2+\sum_{i=1}^{N-1}\left(\frac{y_{m,i}^{\text{exp}}-y_{m,i}^{\text{est}}}{\sigma_{m,i}^y}\right)^2\right], \tag{A.6}$$

where $y_{m,i}^{\text{exp}}$ is the measured mole fraction of species i in the vapour phase of data point m and $\sigma_{m,i}^y$ is its standard deviation. NDG is the number of data groups and NP_n is the number of data points in data group n. Equation (A.6) is derived from Equation (A.5) however, it is to be minimised. For the case studies presented in Chapter 4 the Maximum Likelihood objective function can be expressed in the following simplified form

$$ML = \sum_{k\in K}\left[\left(\frac{T_k^{\text{exp}}-T_k^{\text{est}}}{\sigma_k^T}\right)^2 \sum_{i=1}^{N-1}\sum_{j\in J}\left(\frac{x_{i,j,k}^{\text{exp}}-x_{i,j,k}^{\text{est}}}{\sigma_{i,j,k}^x}\right)^2\right]. \tag{A.7}$$

In the present work, the predictions $\mathbf{x}^{exp}$ are generated by solving the flash problem (Equations (2.9a) - (2.9c)) for a G^E model depending on the vector of estimated parameters $\mathbf{p}$. The optimality conditions of the flash problem and the observed number of phase splits and phases per phase split lead to constraints in the formulation of the parameter estimation.

B. Literature review on binary fluid systems considered in this work

Table B.1.: Literature review on regressions

System	Model	Reference	#Species in the fitted system	#Fits for this system
2,5-DMF/water	NRTL	(20)	3	1
2,5-DMF/water	UNIQUAC	(20)	3	1
2-MF/water	NRTL	(88)	2	2
2-MF/water	UNIQUAC	(88)	2	2
2-MF/water	NRTL	(77)	3	3
2-MF/water	UNIQUAC	(77)	3	3
2-MTHF/water	NRTL	(76)	2	4
2-MTHF/water	UNIQUAC	(76)	2	4
2-MTHF/water	NRTL	(33)	2	2
2-MTHF/water	UNIQUAC	(50)	2	1
butanone/water	NRTL	(76)	2	8
butanone/water	UNIQUAC	(76)	2	8
butanone/water	NRTL	(4)	3	3
butanone/water	UNIQUAC	(4)	3	3
butanone/water	extended UNIQUAC	(91)	2	4
butanone/water	modified UNIQUAC	(91)	2	4
butanone/water	NRTL	(62)	2	1
butanone/water	NRTL	(27)	2	1
butanone/water	NRTL	(46)	3	2
butanone/water	UNIQUAC	(46)	3	2
butanone/water	NRTL	(30)	3	2
butanone/water	UNIQUAC	(30)	3	2

C. Treatment of uncertainties

Note that, in the present work, no distinction is made between the standard uncertainties and standard deviations reported in the literature and thus, both are denoted by u. Uncertainties are rounded to one significant digit unless the leading digit is unity, and the last reported significant digit of mole fractions has the same order of magnitude as the uncertainty, as suggested by Taylor (84). If only the uncertainty of the mole fraction of water is available, the value of the mole fraction of species 1 is rounded to the magnitude of this uncertainty. If no value for the uncertainty of the composition is given, we report with the same number of significant digits as the reference. As some of the mole fractions reported in this chapter are calculated from mass fractions given in the literature, the given uncertainties are converted with the method of propagation of uncertainties that is described by Taylor (84). The equation that results for the uncertainty of a mole fraction is derived in Section C.1. Temperatures are reported in Kelvin with an accuracy of five significant digits. Similarly, regressed parameters are reported with an accuracy of five significant digits. The predicted mole fractions are reported with the same number of significant digits as the measured mole fraction for the respective phase that is reported with the lowest accuracy. However, it is difficult to estimate the difference between predictions of flash problems with rounded and unrounded NRTL parameters, as the predicted mole fractions are not explicit analytical functions of the employed G^E-model.

C.1. Propagation of uncertainties

The usual approach to estimate uncertainties of functions that are calculated from several non-correlated measured values is to apply the simplification that is based on the first-order Taylor series expansion as explained for example by John R. Taylor (84). If the uncertainties of the measured values $y, ..., z$ are independent and random, then the uncertainty of the function

$x(y, ..., z)$ is calculated according to Taylor (84)

$$u(x) = \sqrt{\left(\frac{\partial x}{\partial y} \cdot u(y)\right)^2 + ... + \left(\frac{\partial x}{\partial z} \cdot u(z)\right)^2}. \tag{C.1}$$

In the present work, the mole fraction x is calculated as a function of the mass fraction w and the molar masses M_1, M_2 of species 1 and 2 according to:

$$x = \frac{\frac{w}{M_1}}{\frac{w}{M_1} + \frac{1-w}{M_2}}, \tag{C.2}$$

as the errors of the measured values are assumed to be independent and random.The uncertainty of the mole fraction is hence

$$u(x) = \sqrt{\left(\frac{\partial x}{\partial w} \cdot u(w)\right)^2 + \left(\frac{\partial x}{\partial M_1} \cdot u(M_1)\right)^2 + \left(\frac{\partial x}{\partial M_2} \cdot u(M_2)\right)^2}. \tag{C.3}$$

If the uncertainties of the molar masses M_1 and M_2 are negligible, this results in:

$$u(x) = \frac{u(w)}{M_1 \cdot M_2 \cdot \left(\frac{w}{M_1} + \frac{1-w}{M_2}\right)^2}. \tag{C.4}$$

In the present work, Equation (C.4) is employed to convert uncertainties from mass fractions to mole fractions.

D. LLE measurement data

Table D.1.: Measurement data used for regression of the LLE 2,5-DMF/water.

	Organic phase			Aqueous phase			
T	$x_{2,5-\mathrm{DMF}}$	$u(x_{2,5-\mathrm{DMF}})$	$u(x_{\mathrm{H_2O}})$	$x_{2,5-\mathrm{DMF}}$	$u(x_{2,5-\mathrm{DMF}})$	$u(x_{\mathrm{H_2O}})$	Ref.
K	mol/mol	mol/mol	mol/mol	mol/mol	mol/mol	mol/mol	
283.15	9.961E-01	not reported	2.E-04	1.6E-03	2.E-04	not reported	(21)
293.15	9.960E-01	not reported	5.E-04	1.4E-03	5.E-04	not reported	(21)
303.15	9.952E-01	not reported	2.E-04	1.2E-03	4.E-04	not reported	(21)
298.20	9.95E-01	5.E-03	5.E-03	2.E-04	2.E-04	2.E-04	(20)
313.20	9.92E-01	5.E-03	5.E-03	2.E-04	2.E-04	2.E-04	(20)
298.15	9.921E-01	not reported	9.E-04	not reported	not reported		(25)

Table D.2.: Measurement data used for regression of the LLE 2,5-DMTHF/water.

	Organic phase			Aqueous phase			
T K	$x_{2,5-\mathrm{DMTHF}}$ mol/mol	$u(x_{2,5-\mathrm{DMTHF}})$ mol/mol	$u(x_{\mathrm{H_2O}})$ mol/mol	$x_{2,5-\mathrm{DMTHF}}$ mol/mol	$u(x_{2,5-\mathrm{DMTHF}})$ mol/mol	$u(x_{\mathrm{H_2O}})$ mol/mol	Ref.
283.15	9.23E-01	not reported	1.0E-03	1.51E-02	3.E-04	not reported	(21)
293.15	9.10E-01	not reported	1.5E-03	9.2E-03	2.E-04	not reported	(21)
303.15	8.98E-01	not reported	5.E-03	7.3E-03	1.5E-04	not reported	(21)
273.15	9.08E-01	not reported	3.E-03	2.114E-02	6.E-05	not reported	(81)
282.35	9.06E-01	not reported	3.E-03	1.538E-02	6.E-05	not reported	(81)
292.65	8.99E-01	not reported	3.E-03	1.064E-02	6.E-05	not reported	(81)
304.15	8.93E-01	not reported	3.E-03	7.63E-03	6.E-05	not reported	(81)
313.95	8.87E-01	not reported	3.E-03	5.85E-03	6.E-05	not reported	(81)
323.85	8.83E-01	not reported	3.E-03	4.72E-03	6.E-05	not reported	(81)
334.65	8.52E-01	not reported	3.E-03	4.01E-03	6.E-05	not reported	(81)
344.95	8.68E-01	not reported	3.E-03	3.53E-03	6.E-05	not reported	(81)
356.15	8.68E-01	not reported	3.E-03	2.81E-03	6.E-05	not reported	(81)

Table D.3.: Measurement data used for regression of the LLE 2-MF/water.

	Organic phase			Aqueous phase			
T K	$x_{2-\mathrm{MF}}$ mol/mol	$u(x_{2-\mathrm{MF}})$ mol/mol	$u(x_{\mathrm{H_2O}})$ mol/mol	$x_{2-\mathrm{MF}}$ mol/mol	$u(x_{2-\mathrm{MF}})$ mol/mol	$u(x_{\mathrm{H_2O}})$ mol/mol	Ref.
283.15	9.975E-01	not reported	8.E-04	7.E-03	2.E-03	not reported	(21)
293.15	9.95E-01	not reported	2.E-03	2.E-03	1.0E-03	not reported	(21)
303.15	9.92E-01	not reported	3.E-03	1.7E-03	3.E-04	not reported	(21)
298.15	9.91E-01	not reported		7.E-04	not reported		(73)
298.84	not reported	not reported		7.E-04	1.2E-04	not reported	(88)
312.83	not reported	not reported		7.2E-04	2.E-05	not reported	(88)
333.41	not reported	not reported		7.9E-04	6.E-05	not reported	(88)
353.36	not reported	not reported		1.01E-03	9.E-05	not reported	(88)
374.65	not reported	not reported		1.22E-03	5.E-05	not reported	(88)
393.65	not reported	not reported		1.91E-03	6.E-05	not reported	(88)
413.35	not reported	not reported		2.5E-03	1.3E-04	not reported	(88)
293.15	9.928E-01	not reported		6.4E-04	not reported		(43)
328.15	9.501E-01	not reported		1.8E-03	not reported		(43)

Table D.4.: Measurement data used for regression of the LLE 2-MTHF/water.

	Organic phase			Aqueous phase			
T	$x_{2-\mathrm{MTHF}}$	$u(x_{2-\mathrm{MTHF}})$	$u(x_{\mathrm{H_2O}})$	$x_{2-\mathrm{MTHF}}$	$u(x_{2-\mathrm{MTHF}})$	$u(x_{\mathrm{H_2O}})$	Ref.
K	mol/mol	mol/mol	mol/mol	mol/mol	mol/mol	mol/mol	
283.15	8.470E-01	not reported	2.E-04	5.78E-02	9.E-04	not reported	(21)
293.15	8.350E-01	not reported	3.E-04	3.51E-02	5.E-04	not reported	(21)
303.15	8.2900E-01	not reported	3.E-05	2.82E-02	2.E-04	not reported	(21)
273.15	8.34E-01	not reported	7.E-03	5.27E-02	6.E-04	not reported	(81)
282.65	8.30E-01	not reported	7.E-03	4.33E-02	6.E-04	not reported	(81)
292.45	8.30E-01	not reported	7.E-03	3.40E-02	5.E-04	not reported	(81)
302.65	8.23E-01	not reported	7.E-03	2.62E-02	5.E-04	not reported	(81)
312.75	8.16E-01	not reported	7.E-03	2.08E-02	5.E-04	not reported	(81)
323.25	8.16E-01	not reported	7.E-03	1.74E-02	5.E-04	not reported	(81)
333.85	8.16E-01	not reported	7.E-03	1.46E-02	5.E-04	not reported	(81)
343.75	7.99E-01	not reported	7.E-03	1.32E-02	5.E-04	not reported	(81)
275.15	8.3911E-01	not reported		1.667E-02	not reported		(42)
308.15	8.3706E-01	not reported		8.390E-03	not reported		(42)
323.15	8.3003E-01	not reported		5.953E-03	not reported		(42)
333.15	8.2203E-01	not reported		4.636E-03	not reported		(42)
293.15	8.20E-01	2.E-03	not reported	3.054E-02	6.E-05	not reported	(33)
303.15	8.21E-01	2.E-03	not reported	2.292E-02	2.E-05	not reported	(33)
313.15	8.22E-01	5.E-03	not reported	1.95E-02	9.E-04	not reported	(33)
323.15	8.09E-01	6.E-03	not reported	1.6E-02	2.E-03	not reported	(33)
333.15	8.16E-01	3.E-03	not reported	1.47E-02	6.E-04	not reported	(33)

Table D.4.: Measurement data used for regression of the LLE 2-MTHF/water. (continued)

	Organic phase			Aqueous phase			
T K	$x_{2-\mathrm{MTHF}}$ mol/mol	$u(x_{2-\mathrm{MTHF}})$ mol/mol	$u(x_{\mathrm{H_2O}})$ mol/mol	$x_{2-\mathrm{MTHF}}$ mol/mol	$u(x_{2-\mathrm{MTHF}})$ mol/mol	$u(x_{\mathrm{H_2O}})$ mol/mol	Ref.
343.15	7.95E-01	6.E-03	not reported	1.14E-02	4.E-04	not reported	(33)
293.15	8.478E-01	2.E-04	not reported	2.49E-02	3.E-04	not reported	(33)
303.15	8.33E-01	8.E-03	not reported	2.741E-02	5.E-05	not reported	(33)
333.15	8.2E-01	9.9E-03	not reported	1.18E-02	5.E-04	not reported	(33)
363.15	8.38E-01	6.E-03	not reported	1.173E-02	5.E-05	not reported	(33)
393.15	8.0E-01	2.E-02	not reported	9.5E-03	4.E-04	not reported	(33)
423.15	7.76E-01	2.E-03	not reported	1.23E-02	4.E-04	not reported	(33)
287.20	8.55E-01	8.E-03	2.E-02	3.86E-02	2.E-03	9.E-04	(50)
288.50	8.55E-01	8.E-03	2.E-02	3.83E-02	2.E-03	8.E-04	(50)
293.10	8.54E-01	8.E-03	2.E-02	3.37E-02	2.E-03	8.E-04	(50)
293.60	8.54E-01	8.E-03	2.E-02	3.44E-02	2.E-03	8.E-04	(50)
298.20	8.53E-01	8.E-03	2.E-02	3.00E-02	2.E-03	8.E-04	(50)
298.70	8.53E-01	8.E-03	2.E-02	2.99E-02	2.E-03	8.E-04	(50)
299.40	8.44E-01	8.E-03	2.E-02	2.65E-02	2.E-03	8.E-04	(50)
302.60	8.52E-01	8.E-03	2.E-02	2.68E-02	2.E-03	8.E-04	(50)
303.20	8.52E-01	8.E-03	2.E-02	2.67E-02	2.E-03	8.E-04	(50)
307.60	8.51E-01	8.E-03	2.E-02	2.38E-02	2.E-03	8.E-04	(50)
308.20	8.50E-01	8.E-03	2.E-02	2.39E-02	2.E-03	8.E-04	(50)
312.50	8.49E-01	8.E-03	2.E-02	2.19E-02	2.E-03	8.E-04	(50)
313.20	8.49E-01	8.E-03	2.E-02	2.16E-02	2.E-03	8.E-04	(50)

Table D.4.: Measurement data used for regression of the LLE 2-MTHF/water. (continued)

	Organic phase			Aqueous phase			
T	$x_{2-\mathrm{MTHF}}$	$u(x_{2-\mathrm{MTHF}})$	$u(x_{\mathrm{H_2O}})$	$x_{2-\mathrm{MTHF}}$	$u(x_{2-\mathrm{MTHF}})$	$u(x_{\mathrm{H_2O}})$	Ref.
K	mol/mol	mol/mol	mol/mol	mol/mol	mol/mol	mol/mol	
317.60	8.51E-01	8.E-03	2.E-02	1.95E-02	2.E-03	7.E-04	(50)
318.20	8.50E-01	8.E-03	2.E-02	1.98E-02	2.E-03	7.E-04	(50)
322.10	8.36E-01	8.E-03	2.E-02	1.55E-02	2.E-03	7.E-04	(50)
322.50	8.46E-01	8.E-03	2.E-02	1.78E-02	2.E-03	7.E-04	(50)
323.20	8.46E-01	8.E-03	2.E-02	1.77E-02	2.E-03	7.E-04	(50)
326.90	8.46E-01	8.E-03	2.E-02	1.65E-02	2.E-03	7.E-04	(50)
328.20	8.44E-01	8.E-03	2.E-02	1.65E-02	2.E-03	7.E-04	(50)
332.30	8.42E-01	8.E-03	2.E-02	1.55E-02	2.E-03	7.E-04	(50)
333.20	8.40E-01	8.E-03	2.E-02	1.53E-02	2.E-03	7.E-04	(50)
337.10	8.36E-01	8.E-03	2.E-02	1.44E-02	2.E-03	7.E-04	(50)
338.20	8.35E-01	8.E-03	2.E-02	1.44E-02	2.E-03	7.E-04	(50)
341.10	8.41E-01	8.E-03	2.E-02	1.24E-02	2.E-03	7.E-04	(50)
342.30	8.38E-01	8.E-03	2.E-02	1.36E-02	2.E-03	7.E-04	(50)
343.20	8.36E-01	8.E-03	2.E-02	1.35E-02	2.E-03	7.E-04	(50)

Table D.5.: Measurement data used for regression of the LLE butanone/water.

	Organic phase			Aqueous phase			
T	x_{butanone}	$u(x_{\mathrm{butanone}})$	$u(x_{\mathrm{H_2O}})$	x_{butanone}	$u(x_{\mathrm{butanone}})$	$u(x_{\mathrm{H_2O}})$	Ref.
K	mol/mol	mol/mol	mol/mol	mol/mol	mol/mol	mol/mol	
261.75	6.22E-01	not reported		1.63E-01	not reported		(18)
278.15	6.26E-01	not reported		1.15E-01	not reported		(18)
298.15	6.56E-01	not reported		7.88E-02	not reported		(18)
323.15	6.28E-01	not reported		6.09E-02	not reported		(18)
338.15	6.02E-01	not reported		5.30E-02	not reported		(18)
358.15	5.66E-01	not reported		5.40E-02	not reported		(18)
388.15	4.54E-01	not reported		6.26E-02	not reported		(18)
413.15	2.71E-01	not reported		1.04E-01	not reported		(18)
415.75	upper critical point			1.70E-01	not reported		(18)
273.15	6.71E-01	not reported	3.E-03	1.218E-01	9.E-04	not reported	(81)
282.75	6.67E-01	not reported	3.E-03	1.009E-01	8.E-04	not reported	(81)
292.45	6.65E-01	not reported	3.E-03	8.70E-02	8.E-04	not reported	(81)
302.85	6.62E-01	not reported	3.E-03	7.50E-02	8.E-04	not reported	(81)
312.75	6.53E-01	not reported	3.E-03	6.58E-02	7.E-04	not reported	(81)
322.85	6.49E-01	not reported	3.E-03	6.09E-02	7.E-04	not reported	(81)
333.75	6.18E-01	not reported	3.E-03	5.20E-02	7.E-04	not reported	(81)
343.35	6.11E-01	not reported	3.E-03	5.27E-02	7.E-04	not reported	(81)
263.15	5.937E-01	not reported		1.448E-01	not reported		(30)
273.15	6.216E-01	not reported		1.209E-01	not reported		(30)
313.15	6.601E-01	not reported		5.54E-02	not reported		(30)

Table D.5.: Measurement data used for regression of the LLE butanone/water. (continued)

	Organic phase			Aqueous phase			
T K	x_{butanone} mol/mol	$u(x_{\mathrm{butanone}})$ mol/mol	$u(x_{\mathrm{H_2O}})$ mol/mol	x_{butanone} mol/mol	$u(x_{\mathrm{butanone}})$ mol/mol	$u(x_{\mathrm{H_2O}})$ mol/mol	Ref.
273.15	6.429E-01	not reported		1.679E-01	not reported		(27)
278.15	6.631E-01	not reported		1.403E-01	not reported		(27)
283.15	6.670E-01	not reported		1.158E-01	not reported		(27)
288.15	6.608E-01	not reported		1.027E-01	not reported		(27)
293.15	6.612E-01	not reported		9.30E-02	not reported		(27)
298.15	6.571E-01	not reported		8.62E-02	not reported		(27)
308.15	6.419E-01	not reported		7.18E-02	not reported		(27)
318.15	6.214E-01	not reported		6.50E-02	not reported		(27)
328.15	6.075E-01	not reported		5.42E-02	not reported		(27)
338.15	5.890E-01	not reported		5.18E-02	not reported		(27)
298.15	6.58E-01	2.E-03	2.E-03	7.8E-02	2.E-03	2.E-03	(4)
308.15	6.47E-01	2.E-03	2.E-03	6.6E-02	2.E-03	2.E-03	(4)
318.15	6.43E-01	2.E-03	2.E-03	6.1E-02	2.E-03	2.E-03	(4)
288.15	6.5E-01	2.E-02	2.E-02	9.3E-02	4.E-03	4.E-03	(69)
289.15	6.5E-01	2.E-02	2.E-02	9.2E-02	4.E-03	4.E-03	(69)
290.15	6.5E-01	2.E-02	2.E-02	9.1E-02	4.E-03	4.E-03	(69)
291.15	6.5E-01	2.E-02	2.E-02	8.9E-02	4.E-03	4.E-03	(69)
292.15	6.5E-01	2.E-02	2.E-02	8.7E-02	4.E-03	4.E-03	(69)
293.15	6.5E-01	2.E-02	2.E-02	8.6E-02	4.E-03	4.E-03	(69)
298.15	6.4E-01	2.E-02	2.E-02	7.8E-02	4.E-03	4.E-03	(69)

Table D.5.: Measurement data used for regression of the LLE butanone/water. (continued)

	Organic phase			Aqueous phase			
T K	x_{butanone} mol/mol	$u(x_{\mathrm{butanone}})$ mol/mol	$u(x_{\mathrm{H_2O}})$ mol/mol	x_{butanone} mol/mol	$u(x_{\mathrm{butanone}})$ mol/mol	$u(x_{\mathrm{H_2O}})$ mol/mol	Ref.
303.15	6.3E-01	2.E-02	2.E-02	7.8E-02	4.E-03	4.E-03	(69)
299.85	6.30E-01	not reported		7.12E-02	not reported		(72)
283.15	6.541E-01	not reported		1.0872E-01	not reported		(22)
293.15	6.4058E-01	not reported		9.427E-02	not reported		(22)
303.15	6.2564E-01	not reported		8.906E-02	not reported		(22)

E. Predictions of BOARPET

Table E.1.: BOARPET predictions for the LLE 2,5-DMF/water

	Organic phase	Aqueous phase
Temperature K	$x_{2,5-\mathrm{DMF}}$ mol/mol	$x_{2,5-\mathrm{DMF}}$ mol/mol
283.15	0.996	0.0017
293.15	0.995	0.0010
298.15	0.994	0.0008
298.20	0.994	0.0008
303.15	0.994	0.0007
313.20	0.992	0.0005

Table E.2.: BOARPET predictions for the LLE 2,5-DMTHF/water.

	Organic phase	Aqueous phase
Temperature K	$x_{2,5-\mathrm{DMTHF}}$ mol/mol	$x_{2,5-\mathrm{DMTHF}}$ mol/mol
273.15	0.912	0.0205
282.35	0.908	0.0150
283.15	0.908	0.0146
292.65	0.903	0.0109
293.15	0.903	0.0107
303.15	0.897	0.0080
304.15	0.896	0.0078
313.95	0.890	0.0060
323.85	0.882	0.0048
334.65	0.874	0.0037
344.95	0.865	0.0030
356.15	0.855	0.0024

Table E.3.: BOARPET predictions of Run 1 for the LLE 2-MF/water.

	Organic phase	Aqueous phase
Temperature	$x_{2-\mathrm{MF}}$	$x_{2-\mathrm{MF}}$
K	mol/mol	mol/mol
283.15	0.995	0.004
293.15	0.993	0.002
293.15	0.993	0.002
298.15	0.991	0.002
298.84	0.990	0.002
303.15	0.988	0.002
312.83	0.980	0.001
328.15	0.953	0.001
333.41	0.937	0.001
353.36	0.834	0.001
374.65	0.668	0.001
393.65	0.522	0.001
413.35	0.400	0.002

Table E.4.: BOARPET predictions of Run 2 for the LLE 2-MF/water.

	Organic phase	Aqueous phase
Temperature	$x_{2-\mathrm{MF}}$	$x_{2-\mathrm{MF}}$
K	mol/mol	mol/mol
283.15	0.991	0.002
293.15	0.988	0.002
293.15	0.988	0.002
298.15	0.986	0.002
298.84	0.986	0.002
303.15	0.984	0.002
312.83	0.980	0.002
328.15	0.970	0.002
333.41	0.966	0.002
353.36	0.947	0.001
374.65	0.919	0.001
393.65	0.888	0.001
413.35	0.850	0.001

Table E.5.: BOARPET predictions for the LLE 2-MTHF/water.

T	Organic phase $x_{2-\mathrm{MTHF}}$	Aqueous phase $x_{2-\mathrm{MTHF}}$	T	Organic phase $x_{2-\mathrm{MTHF}}$	Aqueous phase $x_{2-\mathrm{MTHF}}$
K	mol/mol	mol/mol	K	mol/mol	mol/mol
273.15	0.841	0.049	317.60	0.837	0.019
275.15	0.841	0.046	318.20	0.836	0.019
282.65	0.843	0.038	322.10	0.835	0.018
283.15	0.843	0.038	322.50	0.835	0.018
287.20	0.843	0.035	323.15	0.834	0.018
288.50	0.843	0.034	323.20	0.834	0.018
292.45	0.842	0.031	323.25	0.834	0.018
293.10	0.842	0.031	326.90	0.833	0.017
293.15	0.842	0.031	328.20	0.832	0.016
293.60	0.842	0.030	332.30	0.831	0.015
298.20	0.842	0.028	333.15	0.830	0.015
298.70	0.842	0.027	333.20	0.830	0.015
299.40	0.841	0.027	333.85	0.830	0.015
302.60	0.841	0.025	337.10	0.829	0.014
302.65	0.841	0.025	338.20	0.828	0.014
303.15	0.841	0.025	341.10	0.827	0.013
303.20	0.841	0.025	342.30	0.826	0.013
307.60	0.840	0.023	343.15	0.826	0.013
308.15	0.839	0.023	343.20	0.826	0.013
308.20	0.839	0.023	343.75	0.826	0.013
312.50	0.838	0.021	363.15	0.816	0.010
312.75	0.838	0.021	393.15	0.800	0.007
313.15	0.838	0.021	423.15	0.783	0.005
313.20	0.838	0.021			

Table E.6.: BOARPET predictions for the LLE butanone/water.

	Organic phase	**Aqueous phase**		**Organic phase**	**Aqueous phase**
Temperature	x_{butanone}	x_{butanone}	Temperature	x_{butanone}	x_{butanone}
K	mol/mol	mol/mol	K	mol/mol	mol/mol
261.75	0.61	0.169	303.15	0.65	0.077
263.15	0.62	0.163	308.15	0.65	0.072
273.15	0.64	0.128	312.75	0.64	0.069
278.15	0.65	0.115	313.15	0.64	0.068
282.75	0.65	0.106	318.15	0.64	0.065
283.15	0.65	0.105	322.85	0.64	0.062
288.15	0.65	0.096	323.15	0.64	0.062
289.15	0.65	0.095	328.15	0.63	0.059
290.15	0.65	0.093	333.75	0.62	0.057
291.15	0.65	0.092	338.15	0.62	0.055
292.15	0.65	0.090	343.35	0.61	0.053
292.45	0.65	0.090	358.15	0.57	0.051
293.15	0.65	0.089	388.15	0.44	0.058
298.15	0.65	0.082	413.15	0.23	0.113
299.85	0.65	0.081	415.75	0.16	0.162
302.85	0.65	0.077			

F. Comparison of activity coefficients

For a comparison of activity coefficients of 2,5-DMF at infinite dilution in water calculated by Qureshi et al. (64) to those computed from the regressed parameters reported in Subsection 4.3.2, see Table F.1. Qureshi et al. (64) calculate the activity coefficients from an equation based on air-water-partition coefficients, fugacity coefficients, the Poynting correction, the molar volume and the saturation vapour pressure, and report a relative standard uncertainty of 0.02 for the reported activity coefficients. The temperature trend of activity coefficients computed from the regressed NRTL model in the present work is the opposite of the trend reported by Qureshi et al. (64). Considering the lack of measurement data, the regression could also include the activity coefficients at infinite dilution, in addition to mole fraction variables. However, there is no indication that the activity coefficients from literature are determined such that thermodynamic stability can be guaranteed. Consequently, in the present work, we only report these for the sake of completeness.

Table F.1.: Comparison of activity coefficients of 2,5-DMF at infinite dilution in water.

	Activity coefficient	
Temperature	Present report	Qureshi et al. (64)
K	-	-
273.35	381	5583
283.15	624	5664
293.15	989	5523
298.15	1228	5420
303.15	1511	5232
313.15	2230	4817
323.15	3194	4321
333.15	4448	3843

For a comparison of the activity coefficient of 2,5-DMTHF at infinite dilution in water calculated by Cabani et al. (17) to the one computed from the regressed parameters reported in Subsection 4.4.2, see Table F.2. Considering the lack of measurement data, the regression could also include the activity coefficients at infinite dilution, in addition to mole fraction variables. However, there is no indication that the activity coefficients from literature are determined such that thermodynamic stability can be guaranteed. Consequently, in the present work, we only report these for the sake of completeness.

Table F.2.: Comparison of activity coefficients of 2,5-DMTHF at infinite dilution in water.

	Activity coefficient	
Temperature	Present report	Cabani et al. (17)
K	-	-
298.15	124	118.3

For a comparison of activity coefficients of 2-MF at infinite dilution in water calculated by Qureshi et al. (64) to those computed from the regressed parameters reported in Subsection 4.5.2, see Table F.3. Qureshi et al. (64) calculate the activity coefficients from an equation based on air-water-partition coefficients, fugacity coefficients, the Poynting correction, the molar volume and the saturation vapour pressure, and report a relative standard uncertainty of 0.02 for the reported activity coefficients. The temperature trend of activity coefficients computed from the regressed NRTL model in the present work is the opposite of the trend reported by Qureshi et al. (64). Considering the lack of measurement data, the regression could also include the activity coefficients at infinite dilution, in addition to mole fraction variables. However, there is no indication that the activity coefficients from literature are determined such that thermodynamic stability can be guaranteed. Consequently, in the present work, we only report these for the sake of completeness.

Table F.3.: Comparison of activity coefficients of 2-MF at infinite dilution in water.

	Activity coefficient		
Temperature	Run 1	Run 2	Qureshi et al. (64)
K	-	-	-
273.35	143	420	1406
283.15	262	472	1490
293.15	427	522	1507
298.15	522	545	1512
303.15	619	567	1507
313.15	805	605	1398
323.15	946	637	1274
333.15	1015	661	1152

For a comparison of the activity coefficient of 2-MTHF at infinite dilution in water calculated by Cabani et al. (17) to the one computed from the regressed parameters reported in Subsection

4.6.2, see Table F.4. The regression could also include the activity coefficients at infinite dilution, in addition to mole fraction variables. However, there is no indication that the activity coefficients from literature are determined such that thermodynamic stability can be guaranteed. Consequently, in the present work, we only report these for the sake of completeness.

Table F.4.: Comparison of activity coefficients of 2-MTHF at infinite dilution in water.

	Activity coefficient	
Temperature	Present report	Cabani et al. (17)
K	-	-
298.15	52	40.7

G. Aspen Plus DRS results

G.1. LLE 2,5-DMF/water

For the regression in Aspen Plus V8.8 (7) the same measurement data are used as in BOARPET. The Aspen Plus DRS regression tool completes without error messages. The parameters f_{12}^{Aspen}, e_{12}^{Aspen}, c_{12}^{Aspen}, d_{12}^{Aspen}, f_{21}^{Aspen} and e_{21}^{Aspen} are kept constant in the 2,5-DMF/water case study. The regressed NRTL model has the following parameters.

$$
\begin{aligned}
a_{12}^{Aspen} &= 0.20169 && regressed \\
b_{12}^{Aspen} &= 983.53 && regressed \\
f_{12}^{Aspen} &= 0.0000 && fixed \\
e_{12}^{Aspen} &= 0.0000 && fixed \\
c_{12}^{Aspen} &= 0.20000 && fixed \\
d_{12}^{Aspen} &= 0.0000 && fixed \\
a_{21}^{Aspen} &= 62.927 && regressed \\
b_{21}^{Aspen} &= -17204 && regressed \\
f_{21}^{Aspen} &= 0.0000 && fixed \\
e_{21}^{Aspen} &= 0.0000 && fixed
\end{aligned}
$$

The lower bound for the parameter b_{21}^{Aspen} is relaxed to -20000 for the regression of this case study. The predictions for mole fractions and temperatures for the LLE 2,5-DMF/water are presented in Table G.1 as given by Aspen Plus DRS.

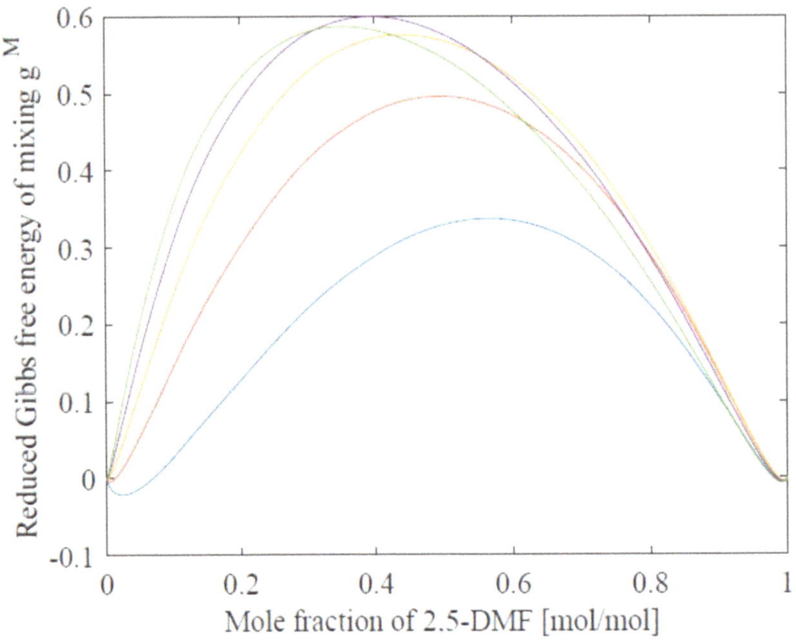

Fig. G.1.: $g^M(x)$ for the LLE 2,5-DMF/water for five equidistant temperatures from 283.15 K to 313.2 K. For all tested temperatures the surface of Gibbs free energy allows for exactly one double tangent.

Table G.1.: Aspen Plus DRS predictions for LLE 2,5-DMF/water.

	Organic phase	**Aqueous phase**
Temperature	$x_{2,5-\mathrm{DMF}}$	$x_{2,5-\mathrm{DMF}}$
K	mol/mol	mol/mol
289.6471	0.9948059	0.00551906
293.2208	0.9949863	0.00260866
295.6332	0.9949488	0.00159906
298.3076	0.9947914	0.00094196
304.4296	0.9940538	0.0002935
307.7546	0.9934561	0.00015944

G.2. LLE 2,5-DMTHF/water

For the regression in Aspen Plus V8.8 (7) the same measurement data are used as in BOARPET. The Aspen Plus DRS regression tool completes without error messages. The parameters f_{12}^{Aspen}, e_{12}^{Aspen}, d_{12}^{Aspen}, f_{21}^{Aspen} and e_{21}^{Aspen} are kept constant in the 2,5-DMTHF/water case study. The regressed NRTL model has the following parameters.

$$
\begin{aligned}
a_{12}^{Aspen} &= -6.5116 && \textit{regressed}\\
b_{12}^{Aspen} &= 1779.2 && \textit{regressed}\\
f_{12}^{Aspen} &= 0.0000 && \textit{fixed}\\
e_{12}^{Aspen} &= 0.0000 && \textit{fixed}\\
c_{12}^{Aspen} &= 0.10000 && \textit{regressed}\\
d_{12}^{Aspen} &= 0.0000 && \textit{fixed}\\
a_{21}^{Aspen} &= 19.189 && \textit{regressed}\\
b_{21}^{Aspen} &= -4128.5 && \textit{regressed}\\
f_{21}^{Aspen} &= 0.0000 && \textit{fixed}\\
e_{21}^{Aspen} &= 0.0000 && \textit{fixed}
\end{aligned}
$$

The parameter c_{12}^{Aspen} is at its lower bound. As the non-randomness parameter α is related to the lattice number in the quasichemical theory of Guggenheim, its bounds should not be selected completely arbitrary, see e.g. Renon et al. (65). The bounds 0.1 and 0.5 are recommended by Pfennig (63). For this reason, the bounds for c_{12}^{Aspen} are not relaxed further. The predictions for mole fractions and temperatures for the LLE 2,5-DMTHF/water are presented in Table G.2 as given by Aspen Plus DRS.

G.3. LLE 2-MF/water

For the regression in Aspen Plus V8.8 (7) the same measurement data are used as in BOARPET. The Aspen Plus DRS regression tool completes without error messages. The parameters f_{12}^{Aspen}, e_{12}^{Aspen}, d_{12}^{Aspen}, f_{21}^{Aspen} and e_{21}^{Aspen} are kept constant in the 2-MF/water case study. The

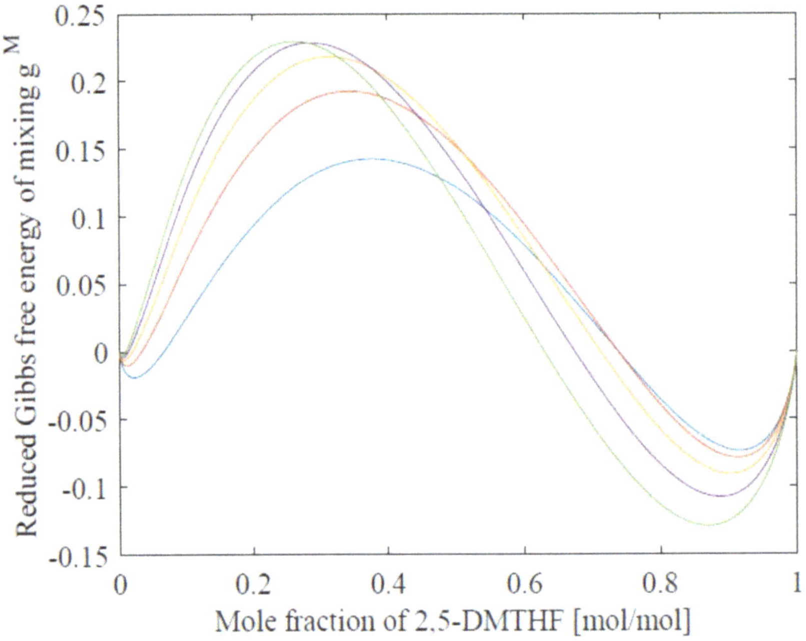

Fig. G.2.: $g^M(x)$ for the LLE 2,5-DMTHF/water for five equidistant temperatures from 273.15 K to 356.15 K. For all tested temperatures the surface of Gibbs free energy allows for exactly one double tangent.

Table G.2.: Aspen Plus DRS predictions for LLE 2,5-DMTHF/water.

	Organic phase	**Aqueous phase**
Temperature	$x_{2,5-\mathrm{DMTHF}}$	$x_{2,5-\mathrm{DMTHF}}$
K	mol/mol	mol/mol
272.9517	0.9097019	0.0200336
282.1821	0.909001	0.0145671
282.9142	0.9088404	0.0142253
292.6622	0.9054973	0.0105668
293.4553	0.9051359	0.0103291
303.3629	0.899651	0.00789906
304.2172	0.8991001	0.00772851
314.0761	0.8919525	0.00608834
323.9331	0.8835064	0.00490336
334.7414	0.8729789	0.00395227
344.7963	0.8621875	0.00329131
355.9981	0.8492474	0.00273094

regressed NRTL model has the following parameters.

$$
\begin{aligned}
a_{12}^{Aspen} &= 7.0491 && regressed\\
b_{12}^{Aspen} &= -1281.7 && regressed\\
f_{12}^{Aspen} &= 0.0000 && fixed\\
e_{12}^{Aspen} &= 0.0000 && fixed\\
c_{12}^{Aspen} &= 0.10000 && regressed\\
d_{12}^{Aspen} &= 0.0000 && fixed\\
a_{21}^{Aspen} &= 1.8863 && regressed\\
b_{21}^{Aspen} &= 810.69 && regressed\\
f_{21}^{Aspen} &= 0.0000 && fixed\\
e_{21}^{Aspen} &= 0.0000 && fixed
\end{aligned}
$$

The parameter c_{12}^{Aspen} is at its lower bound. As the non-randomness parameter α is related to the lattice number in the quasichemical theory of Guggenheim, its bounds should not be selected completely arbitrary, see e.g. Renon et al. (65). The bounds 0.1 and 0.5 are recommended by Pfennig (63). For this reason, the bounds for c_{12}^{Aspen} are not relaxed further. The predictions for mole fractions and temperatures for the LLE 2-MF/water are presented in Table G.3 as given by Aspen Plus DRS.

G.4. LLE 2-MTHF/water

For the regression in Aspen Plus V8.8 (7) the same measurement data are used as in BOAR-PET. The Aspen Plus DRS regression tool completes without error messages. The parameters f_{12}^{Aspen}, e_{12}^{Aspen}, d_{12}^{Aspen}, f_{21}^{Aspen} and e_{21}^{Aspen} are kept constant in the 2-MTHF/water case study.

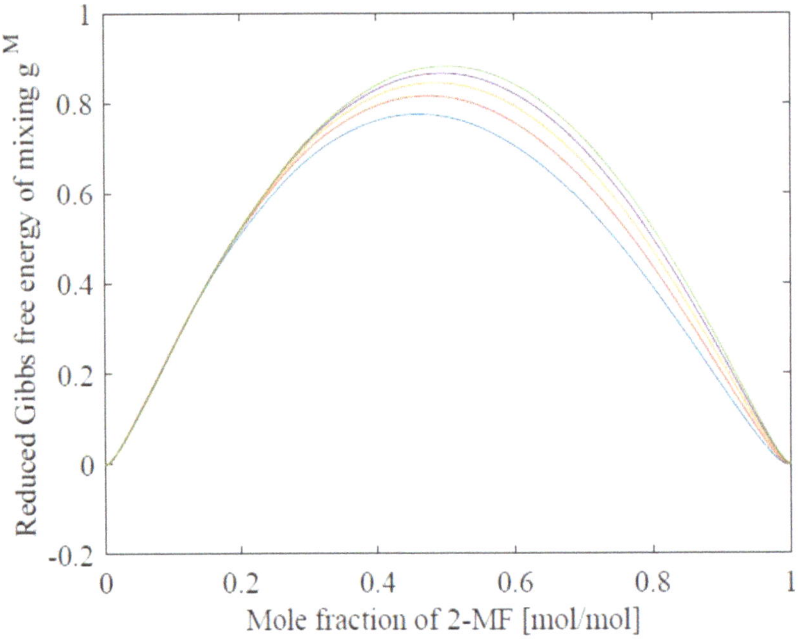

Fig. G.3.: $g^M(x)$ for the LLE 2-MF/water for five equidistant temperatures from 283.15 K to 413.35 K. For all tested temperatures the surface of Gibbs free energy allows for exactly one double tangent.

Table G.3.: Aspen Plus DRS predictions for LLE 2-MF/water.

	Organic phase	**Aqueous phase**
Temperature	$x_{2-\mathrm{MF}}$	$x_{2-\mathrm{MF}}$
K	mol/mol	mol/mol
283.9998	0.9956792	0.00124277
292.3325	0.9961077	0.00125299
292.987	0.9961384	0.00125388
297.0388	0.9963199	0.00125969
300.5494	0.996466	0.00126509
302.3038	0.9965355	0.0012679
314.2264	0.9969528	0.0012888
325.6804	0.9972794	0.00131136
334.3527	0.9974887	0.00132972
353.8926	0.9978694	0.00137417
374.9	0.9981764	0.00142527
393.5902	0.9983875	0.00147253
413.1311	0.9985634	0.00152297

The regressed NRTL model has the following parameters.

$$a_{12}^{Aspen} = 2.5540 \qquad regressed$$
$$b_{12}^{Aspen} = -178.02 \qquad regressed$$
$$f_{12}^{Aspen} = 0.0000 \qquad fixed$$
$$e_{12}^{Aspen} = 0.0000 \qquad fixed$$
$$c_{12}^{Aspen} = 0.38925 \qquad regressed$$
$$d_{12}^{Aspen} = 0.0000 \qquad fixed$$
$$a_{21}^{Aspen} = 7.2683 \qquad regressed$$
$$b_{21}^{Aspen} = -1215.9 \qquad regressed$$
$$f_{21}^{Aspen} = 0.0000 \qquad fixed$$
$$e_{21}^{Aspen} = 0.0000 \qquad fixed$$

The predictions for mole fractions and temperatures for the LLE 2-MTHF/water are presented in Table G.4 as given by Aspen Plus DRS.

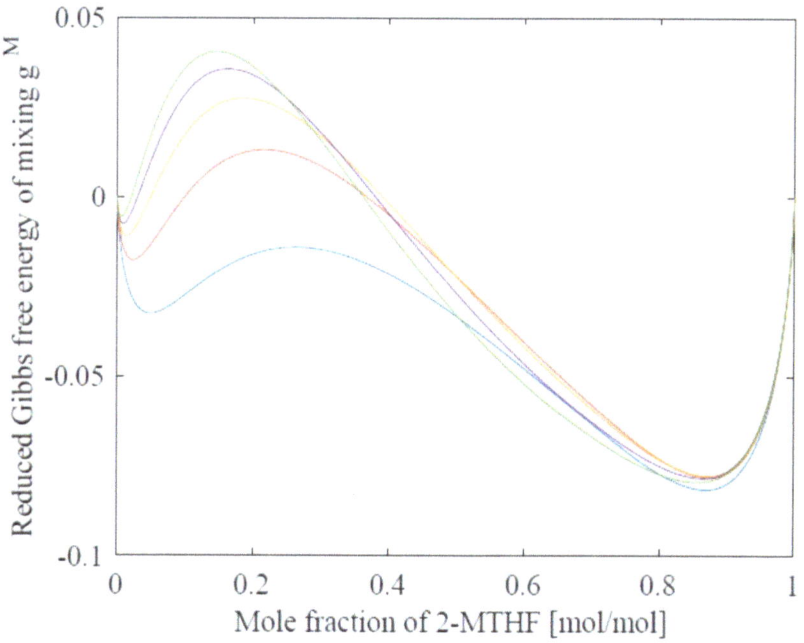

Fig. G.4.: $g^M(x)$ for the LLE 2-MTHF/water for five equidistant temperatures from $273.15\,K$ to $423.15\,K$. For all tested temperatures the surface of Gibbs free energy allows for exactly one double tangent.

Table G.4.: Aspen Plus DRS predictions for LLE 2-MTHF/water.

Temperature K	Organic phase $x_{2-\mathrm{MTHF}}$ mol/mol	Aqueous phase $x_{2-\mathrm{MTHF}}$ mol/mol	Temperature K	Organic phase $x_{2-\mathrm{MTHF}}$ mol/mol	Aqueous phase $x_{2-\mathrm{MTHF}}$ mol/mol
272.7008	0.8401458	0.0432467	313.2045	0.8388795	0.0200152
277.0823	0.8409718	0.0392105	317.5218	0.8378508	0.0187087
282.1557	0.8415593	0.0351875	318.0868	0.8377063	0.0185472
282.2166	0.8415642	0.0351429	322.272	0.8365668	0.0174139
286.8289	0.841793	0.0319934	322.4498	0.8365157	0.0173681
288.103	0.8418102	0.0311963	323.144	0.8363144	0.0171909
292.1433	0.8417434	0.0288526	323.2511	0.836283	0.0171638
292.7631	0.8417175	0.028516	323.2975	0.8362694	0.0171521
292.7858	0.8417165	0.0285039	324.6912	0.8358538	0.0168058
293.0186	0.8417056	0.028379	326.8692	0.8351787	0.0162853
293.2323	0.8416952	0.0282651	328.1471	0.834768	0.0159912
293.3825	0.8416876	0.0281855	332.2574	0.8333757	0.015098
297.9309	0.8413507	0.0259226	333.1531	0.8330579	0.0149134
298.4218	0.8413024	0.0256945	333.189	0.833045	0.0149061
299.3182	0.8412085	0.0252857	333.4803	0.8329405	0.0148469
302.4017	0.8408301	0.0239504	333.8886	0.8327932	0.0147644
302.5003	0.8408166	0.0239095	334.7374	0.8324834	0.0145953
302.8619	0.8407664	0.0237602	337.0726	0.8316077	0.0141448
302.9078	0.84076	0.0237413	338.1558	0.8311898	0.0139431
302.988	0.8407487	0.0237085	341.1877	0.8299813	0.0134012
303.2132	0.8407166	0.0236164	342.2494	0.8295446	0.013219
307.4611	0.8400325	0.0219761	343.1484	0.8291692	0.0130677
308.0366	0.8399285	0.021767	343.3627	0.829079	0.013032
309.6948	0.8396143	0.0211807	343.7675	0.8289079	0.012965
312.3703	0.8390626	0.0202834	362.9755	0.8196008	0.0103089
312.722	0.838986	0.0201697	392.9318	0.8000949	0.00758268
313.0708	0.8389092	0.0200579	422.5586	0.773667	0.00586503

G.5. LLE butanone/water

For the regression in Aspen Plus V8.8 (7) the same measurement data are used as in BOARPET. The Aspen Plus DRS regression tool completes without error messages in the status window, but the following error is recorded in the report file.

- * ERROR: REGRESSION RESULTS DO NOT AGREE WITH 3-PHASE FLASH RESULTS.
  RERUN WITH DIFFERENT PARAMETER INITIAL GUESSES.

The regressed NRTL model has the following parameters.

$$
\begin{aligned}
a_{12}^{Aspen} &= -2.2757 && \textit{regressed}\\
b_{12}^{Aspen} &= 544.77 && \textit{regressed}\\
f_{12}^{Aspen} &= 0.0091321 && \textit{regressed}\\
e_{12}^{Aspen} &= 0.0000 && \textit{fixed}\\
c_{12}^{Aspen} &= 0.42815 && \textit{regressed}\\
d_{12}^{Aspen} &= 0.0000 && \textit{fixed}\\
a_{21}^{Aspen} &= 17.554 && \textit{regressed}\\
b_{21}^{Aspen} &= -2621.8 && \textit{regressed}\\
f_{21}^{Aspen} &= -0.020426 && \textit{regressed}\\
e_{21}^{Aspen} &= 0.0000 && \textit{fixed}
\end{aligned}
$$

The predictions for mole fractions and temperatures for the LLE butanone/water are presented in Table G.5 as given by Aspen Plus DRS. DRS and three-phase flash results are not in agreement for several data points of this case study. Aspen Plus V8.8 (7) prints the predictions for these data points in a history file. The predicted mole fractions from this file are presented in Table G.6.

Figure G.5 shows five plots of reduced Gibbs free energy of mixing over the mole fraction of butanone at equidistant temperatures spanning the complete temperature range of the experiments. As can be seen in Figure G.5, there are multiple possibilities to place double tangents on the surface of Gibbs free energy at elevated temperatures.

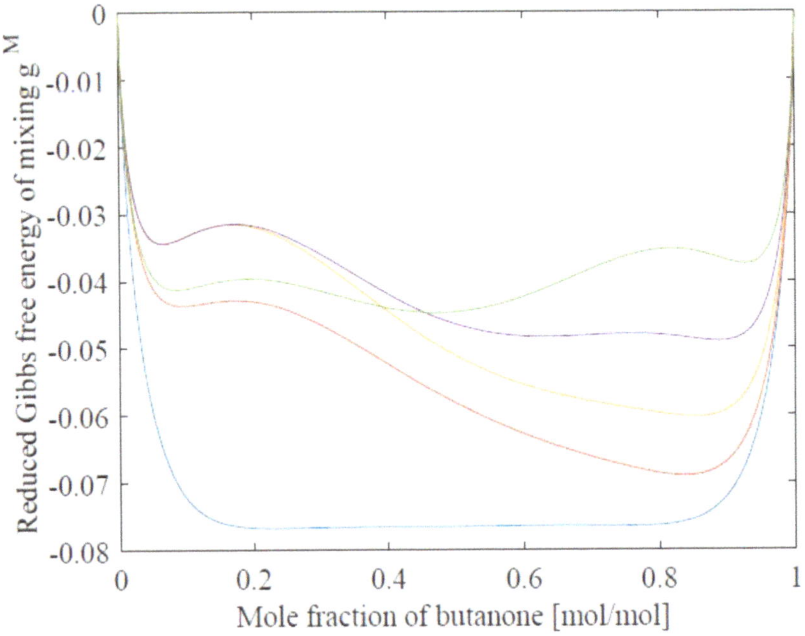

Fig. G.5.: $g^M(x)$ for the LLE butanone/water for five equidistant temperatures from 261.75 K to 415.75 K. In the lower and upper parts of the temperature range of the employed measurement data the surface of Gibbs free energy allows for additional double tangents that do not agree with experimental evidence.

Table G.5.: Aspen Plus DRS predictions for LLE butanone/water.

Temperature K	Organic phase $x_{butanone}$ mol/mol	Aqueous phase $x_{butanone}$ mol/mol	Temperature K	Organic phase $x_{butanone}$ mol/mol	Aqueous phase $x_{butanone}$ mol/mol
263.3389	0.6743397	0.1887958	298.1259	0.6228946	0.0776659
266.5064	0.6093482	0.1636887	299.9197	0.6202931	0.0752647
271.5245	0.6450615	0.1302942	302.5985	0.6123296	0.0667956
273.341	0.6401392	0.1346962	302.6891	0.6113579	0.0725441
273.4653	0.6381463	0.1345779	303.0206	0.616119	0.0717366
277.1634	0.6425759	0.114093	308.0171	0.6074446	0.0675315
278.209	0.6455792	0.1187551	308.1144	0.6064632	0.0676976
282.7157	0.6389281	0.1034166	312.623	0.5969636	0.064468
282.7911	0.6399531	0.1050644	313.1458	0.5904815	0.0630828
283.0201	0.6412157	0.1044602	318.0622	0.59558	0.0610029
287.8171	0.6332932	0.0924317	318.0729	0.5912128	0.0612927
288.1399	0.6355728	0.0932447	322.7093	0.5802528	0.0591116
289.1033	0.6340107	0.0913948	323.0731	0.5872015	0.0588869
290.0963	0.6326793	0.0895781	328.1522	0.5827338	0.0569415
291.1004	0.6313234	0.087841	333.7239	0.5726964	0.0553834
292.0966	0.6299509	0.0861895	338.1276	0.5692527	0.054645
292.3388	0.6276987	0.0858105	338.1573	0.570206	0.0545307
292.8443	0.6262744	0.0838479	343.2967	0.5594284	0.0540222
292.8665	0.6280471	0.0833387	358.1404	0.5436306	0.0538144
293.0759	0.6285555	0.0846126	388.2109	0.5026789	0.0613975
297.8534	0.6192573	0.0767976	414.57	0.1504748	0.0738936
298.0611	0.6205872	0.0777591	415.713	0.178707	0.804483
298.07	0.6201205	0.0777697			

Table G.6.: DRS and three-phase flash results from history file for LLE butanone/water.

	DRS		Three-Phase Flash	
	Organic phase	Aqueous phase	Organic phase	Aqueous phase
Temperature K	x_{butanone} mol/mol	x_{butanone} mol/mol	x_{butanone} mol/mol	x_{butanone} mol/mol
263.34	0.67434	0.188796	0.550391	0.216811
278.21	0.645579	0.118755	0.647246	0.118924
266.51	0.609348	0.163689	0.737538	0.18188
271.52	0.645061	0.130294	0.65781	0.14758
273.34	0.640139	0.134696	0.653395	0.138391
273.47	0.638146	0.134578	0.653184	0.137807
277.16	0.642576	0.114093	0.648404	0.122591
282.72	0.638928	0.103417	0.642254	0.105517
282.79	0.639953	0.105064	0.642168	0.105321
287.82	0.633293	0.0924317	0.636275	0.093879
292.34	0.627699	0.0858105	0.630686	0.0857954
292.84	0.626274	0.0838480	0.630048	0.084996
292.87	0.628047	0.0833387	0.63002	0.084961
293.08	0.628555	0.0846127	0.629755	0.0846356
297.85	0.619257	0.0767977	0.62363	0.077999
298.06	0.620587	0.0777591	0.623361	0.0777419
298.07	0.62012	0.0777698	0.623349	0.0777309
302.6	0.61233	0.0667956	0.617441	0.0726676
302.69	0.611358	0.0725441	0.617322	0.0725761
308.02	0.607445	0.0675316	0.61032	0.067776
308.11	0.606463	0.0676977	0.610192	0.0676983
312.62	0.596964	0.0644681	0.604258	0.0644215
313.15	0.590482	0.0630828	0.60357	0.064081
318.06	0.59558	0.0610029	0.597113	0.061230
318.07	0.591213	0.0612928	0.597099	0.0612242
322.71	0.580253	0.0591116	0.591025	0.0590546
323.07	0.587202	0.0588870	0.590549	0.0589033
328.15	0.582734	0.0569416	0.58391	0.057053
333.72	0.572696	0.0553834	0.576626	0.0555288
338.13	0.569253	0.0546451	0.570852	0.0546558
343.3	0.559428	0.0540223	0.564031	0.0539675
388.21	0.502679	0.0613976	0.497351	0.0612477
414.57	0.150475	0.0738937	0.926629	0.0871688
415.71	0.804483	0.178707	0.491595	0.491595

H. Further formulations for flash simulations

Of course, there are also other reasonable formulations for the simulation of diffusive equilibria in binary fluid systems than those discussed in Chapter 5. For instance, Formulation 5.1 could be reformulated by eliminating the molar phase fractions using Equation (2.9c). The objective function obtained by this elimination is the secant through the predicted compositions evaluated at the feed composition and reads

$$\min_{\mathbf{x}} \left[g^M(x_{1,2}) + \frac{g^M(x_{1,1}) - g^M(x_{1,2})}{x_{1,1} - x_{1,2}} \cdot \left(x_1^0 - x_{1,2}\right) \right]. \tag{H.1}$$

Albeit, Equations (2.9c), (2.9d) and (5.1b) can be removed as constraints after this elimination, the implicit constraint on x_i^0 to lie between $x_{i,1}$ and $x_{i,2}$ has to remain. Therefore,

$$x_{1,2} \leqslant x_1^0 \leqslant x_{1,1}, \tag{H.2}$$

or more generally

$$\left(x_i^0 - x_{i,j}\right)^2 \leqslant \left(x_{i,\hat{j}} - x_{i,j}\right)^2, \text{ for all } i, j, \hat{j} \in \{1,2\}, j \neq \hat{j} \tag{H.3}$$

must be added as constraints. A further kind of objective function can be derived from the Euler equation for extensive Gibbs free energy which reads as follows, see e.g. Stephan et al. (80)

$$nG = \sum_{i \in I} (\mu_i n_i). \tag{H.4}$$

Models of the chemical potentials μ_i usually depend on the compositions of the homogeneous phases in the modelled system. Thus, another objective function can be written as follows, see e.g. Smith et al. (74)

$$\min_{\mathbf{x},\phi} \sum_{j \in J} \phi_j \sum_{i \in I} (\mu_{i,j} x_{i,j}). \tag{H.5}$$

I. Discussion on global solvers

For Formulation 5.3 and parameter set II in combination with the second feed composition from set F_2, BARON reports a solution that is violating the isoactivity condition which, in parts, can be attributed to Formulation 5.3. Another reason is the behaviour of BARON which is described in the following: With the settings specified in Chapter 5, this solution is returned by BARON with the model status "locally optimal" and the solver status "normal completion" after 5 iterations and 0.13 s CPU time. If, however, the absolute termination tolerance is decreased down to 1×10^{-9}, the same solution is returned by BARON with the model status "optimal" and the solver status "normal completion" after 5 iterations and 0.17 s CPU time. The latter absolute termination tolerance is extremely small and may lead to numerical problems, which could be one reason for difficulties with BARON. The secant through the phase compositions that are predicted for the second feed composition in set F_2 is presented in the right diagram in Figure 5.7. The objective function of the last iteration is plotted over the feasible set in Figure I.1. The three maxima of the objective function lie at 0.0117 mol mol^{-1} 2-MTHF with the value 0.00253, 0.583 mol mol^{-1} 2-MTHF with the value zero and 0.937 mol mol^{-1} 2-MTHF with the value 8.66×10^{-5}. If the solver finds the global maximum of Formulation 5.3 in any iteration, then the violation of Baker's criterion for the predictions of the previous iteration is less equal the optimal objective function value plus the optimality tolerance. As the value of the objective function in iteration 5 at $x_{iter} = 0.0117$ mol mol^{-1} is 0.00253 higher than the objective function value of the solution returned by BARON, the author concludes that BARON does not find the global maximum in iteration 5.

Hence, further global solvers are considered for this case. If the same settings as for BARON are used, ANTIGONE predicts the local minimum of the flash problem that is depicted in the middle left diagram in Figure 5.6 after 7 iterations and 2.9 s CPU time. If the same settings are used for LINDOGLOBAL, the termination tolerance of the loop is not reached before $12\,000$ s (≈ 3 h) CPU time. If, however, the tolerance for the violation of Baker's criterion measured in reduced Gibbs free energy is increased to 1×10^{-9}, LINDOGLOBAL finds the global minimum of the flash problem as depicted in the top left diagram in Figure 5.6 after 5 iterations and 4.1 s

CPU time. In a nutshell, one cannot always rely on commercially available "global" solvers of NLP problems to find the global optimum within the specified optimality tolerance. Hence, careful evaluation of the solver results and suitable settings for the solver options are necessary to ensure that a global optimum is found. An exhaustive solver study would be necessary to estimate the reliability of commercially available solvers. However, this is beyond the scope of the present work.

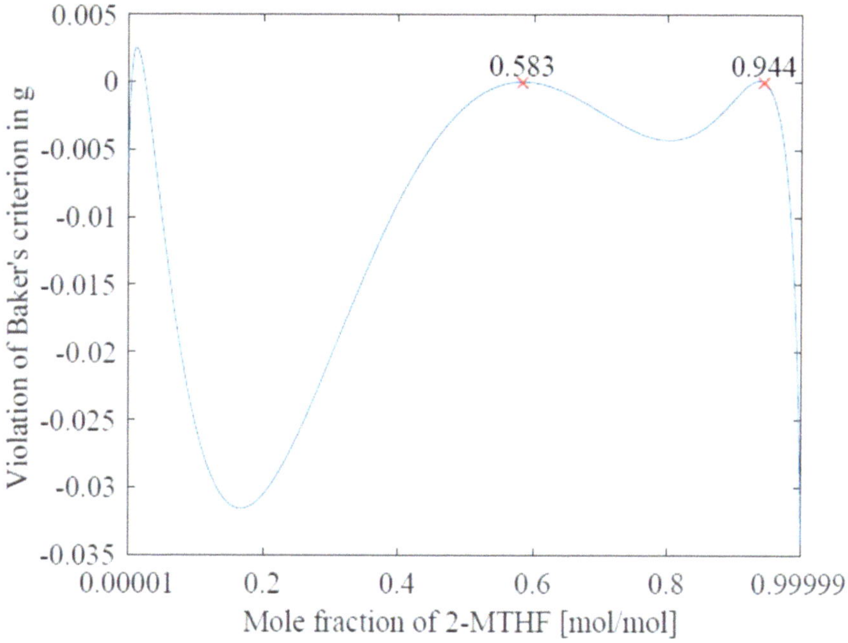

Fig. I.1.: Objective function of Formulation 5.3 in iteration 5 with BARON and parameter set II at $423.176\,\mathrm{K}$ and the second feed composition in set F_2 plotted over the feasible set. BARON returns the local optimum near $0.583\,\mathrm{mol\,mol^{-1}}$ though the global maximum near $0.0117\,\mathrm{mol\,mol^{-1}}$ is desired.

Especially when employing stochastic solvers, Equation (5.4) might be enhanced by only updating the predictions if the objective function value is greater zero. When using deterministic global solvers, Equation (5.4) might be enhanced by only terminating the iterative call of the solver, if the model status after the last iteration is optimal. In case of diffusive equilibria with more than two species, the selection of the phase to be updated is more challenging due to the fact that the new set of predicted phases must be a feasible solution of the flash problem with Formulation 5.1.

J. Example for chemical reaction equilibria

In this chapter, an example given by Tester and Modell (86) is discussed with the aim of stating necessary assumptions for the derivation in this example explicitly. On page 769 in Example 16.5 of their work, the authors present a gas-phase dissociation reaction of elemental iodine to explain the importance of the Gibbs free energy of mixing for chemical equilibria. The reaction reads

$$I_{2(g)} \rightleftharpoons 2\,I_{(g)}. \tag{J.1}$$

The reference states are set to pure vapours with unit fugacity for both substances. The Gibbs free energy minimisation is first discussed assuming ideal mixing and using the following function for the Gibbs free energy of the mixture:

$$\Delta G = \xi \Delta H^0_{rx} - T\xi \Delta S^0_{rx} + RT\left[(1-\xi)\ln\left(\frac{1-\xi}{1+\xi}\right) + 2\xi \ln\left(\frac{2\xi}{1+\xi}\right)\right], \tag{J.2}$$

where ξ denotes the conversion of I_2 to 2 I and ΔH^0_{rx} denotes the enthalpy of reaction for both substances in reference state and ΔS^0_{rx} denotes the entropy of reaction for both substances in reference state. In the next step, gas-phase nonidealities are accounted for by fugacities

$$f_i = y_i \varphi_i, \tag{J.3}$$

where y is the mole fraction in the gas-phase and φ_i is the fugacity coefficient of species i. The symbol γ is used [ibid.] for the fugacity coefficients. However, for clarity of notation φ is used in the present work to distinguish the fugacity coefficient from the activity coefficient. The excess Gibbs free energy is given [ibid.] as

$$\Delta G^E = RT\left[(1-\xi)\ln(\varphi_{I_2}) + 2\xi \ln(\varphi_I)\right]. \tag{J.4}$$

It is stated [ibid.] that Equations J.2 and J.4 together with the description of the example lead to

$$\frac{\partial \Delta G}{\partial \xi} = 0 = \Delta H^0_{rx} - T\Delta S^0_{rx} + RT\ln\left(\frac{\varphi_I^2}{\varphi_{I_2}}\right) + RT\ln\left(\frac{y_I^2}{y_{I_2}}\right) \tag{J.5}$$

in equilibrium. The authors seem to suggest that the right hand side of Equation J.5 is equivalent to the stationarity condition of Gibbs free energy. The Gibbs free energy of the system with gas-phase nonidealities can be expressed as the sum of Equations J.2 and J.4

$$\Delta G = \xi \Delta H^0_{rx} - T\xi \Delta S^0_{rx} + RT\left[(1-\xi)\ln\left(\frac{1-\xi}{1+\xi}\right) + 2\xi \ln\left(\frac{2\xi}{1+\xi}\right)\right] + RT\left[(1-\xi)\ln(\varphi_{I_2}) + 2\xi\ln(\varphi_I)\right]. \quad (J.6)$$

When suppressing the dependence of φ_I and φ_{I_2} on ξ in notation, the derivative of Equation J.6 with respect to the conversion ξ at constant temperature reads

$$\frac{\partial \Delta G}{\partial \xi} = \Delta H^0_{rx} - T\Delta S^0_{rx} - RT\ln\left(\frac{1-\xi}{1+\xi}\varphi_{I_2}\right) + RT(1+\xi)\varphi_{I_2}\left(\frac{\partial \varphi_{I_2}}{\partial \xi}\frac{1-\xi}{1+\xi} - \frac{2\varphi_{I_2}}{(1+\xi)^2}\right) + 2RT\ln\left(\frac{2\xi}{1+\xi}\varphi_I\right) + RT(1+\xi)\varphi_I\left(\frac{\partial \varphi_I}{\partial \xi}\frac{2\xi}{1+\xi} + \frac{2\varphi_I}{(1+\xi)^2}\right). \quad (J.7)$$

It is not apparent in which cases the right hand sides of Equation J.7 and Equation J.5 are equivalent. To clarify this, we simplify Equation J.7 by assuming that both fugacity coefficients are independent of the conversion. In this case Equation J.7 can be expressed as

$$\frac{\partial \Delta G}{\partial \xi} = \Delta H^0_{rx} - T\Delta S^0_{rx} + RT\left(\frac{-2\varphi^2_{I_2}}{1+\xi} + \frac{2\varphi^2_I}{1+\xi} + \ln\left(\frac{y^2_I\varphi^2_I}{y_{I_2}\varphi_{I_2}}\right)\right). \quad (J.8)$$

Still, the right hand side of Equation J.8 is only equivalent to the right hand side of Equation J.5 if the fugacity coefficients of both species are identical. However, it is not stated [ibid.] that the fugacity coefficients of both species are assumed to be equal and that their value is assumed to be independent of the conversion. Note that these assumptions do not always hold. Consequently, it is not clear on which assumptions the authors based their derivation of Equation (J.5). In the author's opinion, assumption that the fugacity coefficients of both species are assumed to be equal and that their value is assumed to be independent of the conversion are not valid because I and I_2 are two different species. Thus, the full model according to Equation (J.5) is not recommended.